Charles Seale-Hayne Library
University of Plymouth
(01752) 588 588
LibraryandITenquiries@plymouth.ac.uk

Ocean–Atmosphere Interactions

Ocean Sciences Research (OSR) Volume 3

Editor: M. M. Takahashi (University of Tokyo)

Editorial Board:

The titles in this series are listed at the end of this volume.

Cover illustration: Flow visualization of breaking wind waves. The wind direction is from left to right. Blue and red stroboscopic light was applied with a time lag of 5 milliseconds at Tohoku University wind-wave tank. Photograph by M. Koga. (cf. *Tellus*, 1981.)

Ocean–Atmosphere Interactions

Edited by

Yoshiaki TOBA

Professor Emeritus of Tohoku University,
Sendai, Japan

Earth Observation Research Center (EORC),
National Space Development Agency of Japan,
Tokyo, Japan

Research Institute for Environmental Sciences and
Public Health of Iwate Prefecture (I-RIEP),
Morioka, Japan

Terra Scientific Publishing Company, Tokyo

Kluwer Academic Publishers, Dordrecht, London, Boston

C.I.P. Catalogue record for this book is available from the Library of Congress.

ISBN 1-4020-1171-7 (Kluwer)

Published by Terra Scientific Publishing Company, 2003 Sansei Jiyugaoka Haimu, 27-19 Okusawa 5-chome, Setagaya-ku, Tokyo 158-0083, Japan / Kluwer Academic Publishers, P.O. Box 17, 3300 AA Dordrecht, The Netherlands.

Kluwer Academic Publishers incorporates
the publishing programmes of
D. Reidel, Martinus Nijhoff, Dr W. Junk and MTP Press.

Sold and Distributed in the U.S.A. and Canada
by Kluwer Academic Publishers,
101 Philip Drive, Assinippi Park, Norwell, MA 02061, U.S.A.
in Japan by Terra Scientific Publishing Company (TERRAPUB),
2003 Sansei Jiyugaoka Haimu, 27-19 Okusawa 5-chome, Setagaya-ku,
Tokyo 158-0083, Japan.

In all other countries, sold and distributed
by Kluwer Academic Publishers,
P.O. Box 322, 3300 AH Dordrecht, The Netherlands.

(This book is partly supported by Grant-in-Aid for Publication Scientific Research Results of the Japan Society for the Promotion of Science.)

Printed in Japan

Contents

Foreword vii

Preface to the English Edition ix

List of Authors xi

Chapter 1. Introduction
  Y. Toba 1

Chapter 2. Air–Sea Interface Processes and Wind Waves
  Y. Toba 13

Chapter 3. Surface Mixed Layer in the Ocean and Water Mass Analysis
  K. Hanawa and T. Suga 63

Chapter 4. Large Scale Ocean–Atmosphere Interactions
  K. Hanawa 111

Chapter 5. Fundamentals of Large-Scale Interaction
  T. Yamagata and Y. Wakata 143

Chapter 6. Numerical Modeling for Large-Scale Ocean–Atmosphere Interactions
  M. Endoh, Y. Kitamura, H. Ishizaki and T. Motoi 195

Chapter 7. Satellite Remote Sensing of the Air–Sea Interaction
  H. Kawamura 239

Index 299

Foreword

The earth's atmosphere and oceans comprise, to our immediate senses, different components of the planet on which we live and thus were in earlier times looked upon as different subjects of study and research. In about the 1960's, a concept was established that the atmosphere and the oceans interact with each other in one system, and the terms "air–sea system" and "ocean–atmosphere system" came into popular use. Further, since the 1980s, these terms have been expanded to link our entire climate system, including the cryosphere that exists over continents and polar regions. On the other hand, the crucial role of the oceans in determining the deposition of anthropogenic carbon dioxide in the atmosphere has been widely recognized. We now live in an age when interdisciplinary research on the ocean and atmosphere, including physical, chemical and biological processes, elucidates the circulation of a wide variety of materials over the earth.

For students of the atmospheric and oceanic sciences, the curriculum of most universities now includes mathematics, physics, chemistry, biology, and languages as essential tools for their study and future research. This book has been written with the curriculum in physical sciences in mind, focusing on interactions between the oceans and atmosphere and including most of the fundamental processes in the climate system. However, we intend for this book to be useful for students, researchers, and field technicians in various research areas and have therefore accommodated up-to-date information such as statistical data, description of various physical phenomena, theories, recent techniques for analyses of observational data, numerical modeling, and so forth.

We now face the urgent task of understanding change and variability of our global environment in relation to human activities and of pursuing its predictability based on a fundamental understanding of the climate system. In Japan, for example, we had in 1993 an unusually severe "yamase" phenomenon, when a cool summer in northeastern Japan caused social problems through a very bad harvest of rice. On the contrary, in the following summer, 1994, we experienced the hottest summer in the history of observation in Japan, and we had some regions where a long-lasting scarcity of rain prevented a regular supply of water to the cities. There is no evidence that these large-amplitude climatological fluctuations were caused by merely stochastic variability. On the other hand, the prediction of a three degree temperature rise within some foreseeable future years does not necessarily mean a uniform warming of the earth. Unusual weather is anticipated in various areas of the earth. We cannot say that the above-mentioned recent unusual climate in Japan was not a manifestation of the global change. But we should continue our research activities. It is our hope that this book will be useful as a textbook for students who wish to enter into research in these areas and a reference book for those working in various related fields.

The seven chapters of this book have been written by a rather small number of colleagues. In the editing process, we tried to assure that the individual chapters comprised a unified textbook, at the same time respecting each author's autonomy and allowing for each chapter to be read separately. The editor now takes this opportunity to express his warmest thanks to my colleagues who wrote those chapters, and to the publisher, The University of Tokyo Press, especially to two colleagues, Ms. Megumi Shimizu and Ms. Rika Tannai who made large efforts in detailed technical editing as well as in planning.

January 1996

Yoshiaki Toba

Preface to the English Edition

This book is an updated English edition of a Japanese book of the same title, edited by myself and published by The University of Tokyo Press in 1996. It is a textbook for graduate courses, comprising up-to-date reviews of the field, written by Japanese scientists including my friends and students. Some days after it was published I was greatly honored to learn from some of my American friends that the contents of this book were used in their classes. It is my hope that this book becomes widely utilized as a textbook for the study of ocean–atmosphere interactions, in an age when global change is an important issue for mankind.

The publication of this English edition is a product of the enthusiastic efforts of Mr. Keiji Oshida and his colleagues, especially Ms. Yumi Terashima and Ms. Yuko Cho, of Terra Scientific Publishing Company (TERRAPUB) together with Wolters-Kluwer Academic Publishers. The Ministry of Education, Science and Culture of Japan supported this publication financially. I am profoundly grateful to all those people who have been concerned with this publication.

Dr. Motoyasu Miyata and Dr. Takuji Waseda of Frontier Research System for Global Change and International Pacific Research Center, University of Hawaii, helped us with the draft translation of the original Japanese edition. Quotations from the Bible, some lines of English poetry and quotations from scientists are supplied courtesy of Dr. Motoyasu Miyata and Professor Akiko Fukuchi of Tohoku Gakuin University. Without their collaboration this publication would not have been successful. I sincerely thank them all for their kind help.

August 2002

Yoshiaki Toba

List of Authors

Chapter 1 and Chapter 2

Professor Yoshiaki Toba[1,2,3]
[1]Professor Emeritus of Tohoku University,
6-7-11 Yagiyama-Minami, Taihaku-ku, Sendai 982-0807, Japan
E-mail: toba@pol.geophys.tohoku.ac.jp
[2]Earth Observation Research Center (EORC),
National Space Development Agency of Japan, Tokyo 104-6023, Japan
[3]Research Institute for Environmental Sciences and Public Health of
Iwate Prefecture (I-RIEP), Morioka 020-0850, Japan

Chapter 3   Professor Kimio Hanawa[1] and Associate Professor Toshio Suga[2]
[1]Department of Geophysics, Graduate School of Science,
Tohoku University, Aoba-ku, Sendai 980-8578, Japan
E-mail: hanawa@pol.geophys.tohoku.ac.jp
[2]Department of Geophysics, Graduate School of Science,
Tohoku University, Aoba-ku, Sendai 980-8578, Japan
E-mail: suga@pol.geophys.tohoku.ac.jp

Chapter 4   Professor Kimio Hanawa
Department of Geophysics, Graduate School of Science,
Tohoku University, Aoba-ku, Sendai 980-8578, Japan
E-mail: hanawa@pol.geophys.tohoku.ac.jp

Chapter 5   Professor Toshio Yamagata[1] and Professor Yoshinobu Wakata[2]
[1]Department of Earth and Planetary Science,
Graduate School of Science, The University of Tokyo,
7-3-1 Hongo, Bunkyo-ku, Tokyo 113-0033, Japan
E-mail: yamagata@eps.s.u-tokyo.ac.jp
[2]Research Institute for Applied Mechanics (RIAM), Kyushu University,
6-1 Kasuga-Kohen, Kasuga-shi, Fukuoka 816-8580, Japan
E-mail: wakata@riam.kyushyu-u.ac.jp

Chapter 6   Dr Masahiro Endoh[1], Dr Yoshiteru Kitamura[2],
               Dr Hiroshi Ishizaki[3] and Dr Tatsuo Motoi[4]
               [1]Center for Climate System Research (CCSR), University of Tokyo,
               Komaba 4-6-1, Meguro-ku, Tokyo 153-8904, Japan
               E-mail: endoh@ccsr.u-tokyo.ac.jp
               [2]Climate and Marine Department, Japan Meteorological Agency,
               Ootemachi 1-3-4, Chiyoda-ku, Tokyo 100-8122, Japan
               E-mail: ykitamur@met.kishou.go.jp
               [3]Oceanographic Research Department, Meteorological Research Institute,
               Nagamine 1-1, Tsukuba, Ibaraki 305-0052, Japan
               E-mail: hishizak@mri-jma.go.jp
               [4]Frontier Research System for Global Change (FRSGC),
               Yokohama Institute for Earth Sciences,
               3173-25, Showa-machi, Kanazawa-ku, Yokohama,
               Kanagawa 236-0001, Japan
               E-mail: tmotoi@jamstec.go.jp

Chapter 7   Professor Hiroshi Kawamura
               Center for Atmospheric and Oceanic Studies (CAOS),
               Graduate School of Science, Tohoku University,
               Aoba-ku, Sendai 980-8578, Japan
               E-mail: kamu@ocean.caos.tohoku.ac.jp

Chapter 1

Introduction

Yoshiaki TOBA

Ocean–Atmosphere Interactions, Ed. Y. Toba, pp. 1–12.
© by TERRAPUB / Kluwer, 2003.

Introduction

Yoshiaki TOBA

And God said, Let the waters under the
heaven be gathered together unto one
place, and let the dry *land* appear: ...
And God called the dry *land* Earth; and
the gathering together of the waters
called he Seas: ...

—*Genesis. 1: 9–10*

1.1 INTRODUCTION TO OCEAN–ATMOSPHERE INTERACTIONS

1.1.1 Energy flow

The main source of the energy that generates various phenomena in the atmosphere and the oceans is radiation from the sun. About 70% of the total solar radiation that reaches the earth is absorbed by the surface of the earth, including the atmosphere and oceans, and the remaining 30% is reflected. Meanwhile, the earth emits energy as thermal radiation corresponding to the temperature of its surface. The balance between the incoming and outgoing energies is in equilibrium at an average temperature of 15°C, after the "greenhouse effect" is taken into account. The fact that liquid water exists on a major part of the earth's surface, in the form of ice at the polar regions and on high mountains, is the result of this balance. It is determined primarily by the distance between the earth and the sun.

Thermal energy coming from the solid earth amounts only to 0.1% of that which enters or exits the ocean and the atmosphere. The ocean and the atmosphere with their motions can thus be considered as a heat engine which is operated by solar radiation.

Though it contains small amounts of chemical constituents, sea water consists predominantly (96.5%) of pure water. For meteorological and oceanographic processes, it is of special importance that water assumes a unique position among other substances in nature (see Table 1-1). For instance, its freezing point is at a normal temperature, and its boiling point is a little higher, allowing its presence in a liquid form covering the earth. Its very high heat capacity enables water to store great quantities of heat without causing large variations of temperature on the earth's surface. The latent heat of fusion is so great that sea water plays the function of a thermostat near the freezing point. Its

Table 1-1. Some unique characteristics of water (partly after Sverdrup *et al.*, 1942).

Characteristics	Comparison with other liquids	Significance in nature
Boiling/freezing point	Especially high among hydrogen compounds of 6b element group	Exists in a liquid form at normal temperature
Specific heat	Largest except NH_3	Makes variations of surface temperature small Heat transport by water circulation is large Keeps body temperature constant
Latent heat of fusion	Largest except NH_3	Works as a thermostat near 0°C
Latent heat of evaporation	Largest	Important for exchange and transport of heat in the ocean–atmosphere system
Infrared absorption	Strong for almost all wavelengths	Important for heat processes in the ocean–atmosphere system
Thermal expansion	Very small	Density variation of sea water is kept small, so that currents cannot be very strong
Compressibility	Very small	Same as above
Surface tension	Largest except mercury	Important in physiology of cells, formation of clouds and rain
Dissociation power	Very large	Ocean is a vast storage of elements

high latent heat of evaporation results in increasing heat exchange between the ocean and the atmosphere when sea water evaporates, and in increasing heat transport within the atmosphere through the water vapor transport. These unique properties play significant roles in motions in the oceans and the atmosphere, their interactions on the large scale, and in the earth's climate system itself. Also, its strong dissociation power and its capacity to dissolve many substances cause materials to circulate in large quantities together with oceanic circulation.

Since the earth is spherical, lower latitudes receive more shortwave solar radiation energy than higher latitudes. This incoming heat energy is made even to some extent in the latitudinal direction due to oceanic and atmospheric motions, and is emitted back to space as long-wave radiation. The difference of the incoming and outgoing radiation between high and low latitudes corresponds to the heat transport from low to high latitudes as a whole, by large-scale oceanic and atmospheric motions.

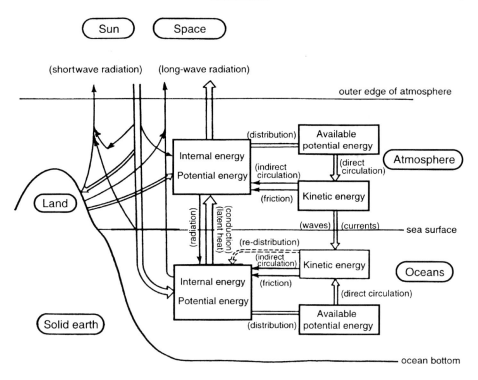

Fig. 1-1. Schematic chart of main energy flow of the atmosphere–ocean system (after Toba, 1970).

Figure 1-1 shows the main energy flow in the ocean–atmosphere system. As is seen, the energy flows back and forth between the ocean and the atmosphere many times. Most of the incoming shortwave radiation energy from the sun passes through the atmosphere and is absorbed by the surface of the earth, particularly by sea water which occupies 72% of the earth's surface. The heat, once absorbed by the sea, is transferred by ocean currents, thus forming a surface temperature distribution of the ocean. The ocean surface warms the atmosphere from underneath according to the temperature difference distributions. The warmed atmosphere generates winds on large scales, which in turn generate currents in the upper ocean. Ocean currents again transport heat and warm the atmosphere according to their temperature distribution.

Since the temperature of the earth's surface is near the freezing point of water, the phase change easily occurs among water vapor, water and ice, causing large amounts of snow and ice to rise and fall over the continents as well, which again influences the energy flow quite significantly. Thus the oceans and the atmosphere operate as a single system of fluids, where water acts as a mediator.

1.1.2 General circulation of oceans and atmosphere

The atmosphere covers the entire earth. The general motions of the atmosphere are divided into three meridional circulation patterns on the average, determined by the angular velocity of the rotation of the earth and the temperature difference between the equator and poles. The direct circulation breaks down into the Hadley cell of low latitudes including extremely persistent Trade Winds, and polar cells in high latitudes, with meandering westerlies in the mid latitudes. Such an atmospheric circulation effectively transports heat from low to high latitudes.

The ocean is also a single body of water covering the earth, except that continents separate it into oceans and seas. The global pattern of heat transport from low to high latitudes differs greatly depending on the locality. For example, patterns in the Pacific and the Atlantic are entirely different from each other.

The oceanic circulation in the surface layer, such as the subtropical circulation which includes the Kuroshio or the Gulf Stream as a western boundary current, is primarily driven by the wind system in the global atmospheric circulation system. Sea water in the North Pacific is considered to make a complete turn in about six years by this surface circulation. On the other hand, the primary cause of the three-dimensional global ocean circulation that reaches the deep layer is considered to be a thermohaline circulation, which includes vertical convection due to density differences. Surface water in high latitudes, having high salinity and cooled by cold strong winds, starts to sink when its density becomes higher than that of water below. Such subduction occurs on large scales in two regions of the oceans: one is southeast of Greenland, the other the Weddell Sea in the Antarctic. The deep water formed near Greenland flows along the western side of the Atlantic, crosses the equator, meets the deep water from the Weddell Sea and flows together to the east in the Antarctic Circumpolar Sea, then flow into the Pacific Ocean. It is estimated that such a thermohaline circulation completes a cycle on the order of 2000 years.

Part of the old deep water that flows into the North Pacific gradually comes up in a broad region and mixes with surface water. Part of the surface water flows from the Indonesian Archipelago to the Indian Ocean to the Atlantic, through waters south of Cape Town and joins the Gulf Stream, returning to offshore Greenland again (Broecker's Conveyer Belt). However, its details need to be studied further. Fundamentals of research by modeling the ocean–atmosphere interactions, including deep circulation, will be described in Chapter 6.

1.1.3 Interactions on various scales

The oceanic general circulation in the surface layer, or the thermohaline deep circulation described above, shows an aspect of large scale ocean–atmosphere interactions. The three dimensional oceanic circulation involves more complexities, such as inter-decadal variations including subduction in mid and high latitudes into the main thermocline, which should be studied further from a viewpoint of detailed ocean–atmosphere interactions. This topic is related to the contents of Chapter 4.

A phenomenon predominantly governed by a direct interaction of the ocean and atmosphere occurs at the equator where Coriolis force on the rotating earth tends to zero. It is ENSO, that is, El Niño in the ocean combined with Southern Oscillation in the atmosphere. ENSO is a phenomenon in the tropical Pacific, but recent studies show that it influences mid and high latitude oceans, the weather in Japan and in the US, something like the heartbeat of the global ocean–atmosphere system (teleconnection). Theoretical aspects of the ENSO will be the main theme of Chapter 5; its observational aspects or the methods of analysis will be described in detail in Chapters 3 and 4.

Increase of carbon dioxide in the atmosphere due to human activities, and global warming as a result, are now a worldwide problem under study. This question is closely related to the exchange of carbon dioxide between the oceans and the atmosphere and with the carbon cycle in the ocean, including biological processes. Understanding of mid and deep circulations in the oceans will be essential for predicting global warming and the resultant sea level rise.

Interactions between the oceans and the atmosphere take place directly through exchange of momentum, energy and materials between the oceans and the atmosphere at the sea surface. This process is carried out by wind waves, which are caused by the wind blowing over the sea surface. We have some empirical laws on wind waves in the sea, and forecasting their generation and growth is possible to a certain accuracy. However, to elucidate them in hydrodynamic context is an extremely difficult task, and we haven't yet achieved enough theoretical understanding. And yet the researches of wind waves are being undertaken internationally as contemporary questions, in relation to carbon dioxide exchange between the oceans and the atmosphere under strong winds, as well as to satellite observations of the sea. These topics will be treated in Chapter 2.

Progress during the past decade with satellite observations has made it possible for their sea surface data to be effectively applied. Microwave sensors can measure waves on the sea surface under clouds as well as distributions of wind over the sea surface, based on information about high frequency components of wind waves. It is expected that these data will help to understand the mechanisms of ocean–atmosphere interactions. They will also be utilized for weather forecasting and numerical modeling to track and predict variations of the oceans and the atmosphere. This theme will be developed in Chapter 7.

1.2 FUNDAMENTAL EQUATIONS OF FLUID MOTIONS AND NONDIMENSIONALIZATION

1.2.1 Equation of conservation of physical quantities

Consider an infinitesimal volume ΔV containing a group of fluid particles. Let G be a physical quantity per unit volume; then the material derivative of the physical quantity $G\Delta V$ contained in ΔV is written as

$$\frac{D}{Dt}(G\Delta V) = K\Delta V, \tag{1-1}$$

where K is the rate of generation, or the rate of inflow by diffusion, of the physical quantity per unit volume per unit time. This, with v the velocity vector, leads to the following *equation of conservation* for a physical quantity:

$$\frac{DG}{Dt} + G\nabla \bullet v = K \tag{1-2}$$

or

$$\frac{\partial G}{\partial t} + \nabla \bullet (Gv) = K. \tag{1-3}$$

G can be mass, momentum, kinetic energy, thermal energy, salinity, vorticity, tracer element concentration, etc.

1.2.2 Equation of continuity

If we take density ρ as G and assume $K = 0$, Eq. (1-3) becomes the *equation of continuity*

$$\frac{\partial \rho}{\partial t} + \nabla \bullet (\rho v) = 0. \tag{1-4}$$

When $D\rho/Dt = 0$, we obtain the simplest form of the *equation of continuity for an incompressible fluid*,

$$\nabla \bullet v = 0. \tag{1-5}$$

1.2.3 Equation of motion

If we take the momentum ρv as G and the force as K, we obtain the equation of motion. If the force contains only pressure p, it becomes *Euler's equation of motion*, but since both the oceans and the atmosphere are generally of turbulent motion, we should take into account the effect of turbulence. To begin with, replacing G in Eq. (1-2) with ρv, we can rewrite the equation in a component form:

$$\frac{\partial(\rho v_i)}{\partial t} + \frac{\partial}{\partial x_k}(\rho v_i v_k) = -\frac{\partial p}{\partial x_i} + \rho F_i \tag{1-6}$$

where $\boldsymbol{F}$ is the external force acting on unit mass.
      This can be rewritten as

$$\frac{\partial(\rho v_i)}{\partial t} + \frac{\partial \pi_{ik}}{\partial x_k} = \rho F_i , \tag{1-7}$$

$$\pi_{ik} = p\delta_{ik} + \rho v_i v_k \tag{1-8}$$

where δ_{ik} is the Kronecker delta

$$\delta_{ik} = \begin{cases} 1, & i = k \\ 0, & i \neq k . \end{cases}$$

The tensor π_{ik} is called the *momentum flux density*.
      For a viscous fluid, Eq. (1-7) will have an additional term of *viscous stress tensor*,

$$\sigma_{ik} = \mu \left(\frac{\partial v_i}{\partial x_k} + \frac{\partial v_k}{\partial x_i} \right) \tag{1-9}$$

and Eq. (1-8) becomes

$$\pi_{ik} = \rho v_i v_k + p\delta_{ik} - \sigma_{ik} \tag{1-10}$$

where μ is the *coefficient of viscosity*. Then Eq. (1-7) is another form of the *Navier–Stokes equation (N–S equation)*

$$\frac{Dv}{Dt} = -\frac{1}{\rho}\nabla p + v\nabla^2 v + \boldsymbol{F} , \tag{1-11}$$

where $v = \mu/\rho$ is the *kinematic coefficient of viscosity*.

1.2.4 Equations extended to turbulence conditions

      In a turbulent motion let v_i denote the velocity component, which is composed of a mean velocity $\overline{v_i}$ on which random fluctuations v_i' are superimposed:

$$v_i = \overline{v_i} + v_i' . \tag{1-12}$$

Then, using Reynolds' axiom[1], the mean momentum flux density τ_{ik} becomes

$$\overline{\pi_{ik}} = \rho\overline{v_i v_k} + p\overline{\delta_{ik}} - \overline{\sigma_{ik}} + \rho\overline{v_i' v_k'}, \tag{1-13}$$

$$\overline{\sigma_{ik}} = \mu\left(\frac{\partial\overline{v_i}}{\partial x_k} + \frac{\partial\overline{v_k}}{\partial x_i}\right). \tag{1-14}$$

That is, in addition to the viscous stress tensor for the momentum flow, we have now the *eddy viscosity* or the *Reynolds stress* $-\rho\overline{v_i' v_k'}$.

The actual form of the Reynolds stress is determined by the structure of turbulence pertinent to the phenomenon under consideration. However, in analogy to the molecular viscous term, assuming that this stress is proportional to the shear of the mean velocity, let us write, for example,

$$-\rho\overline{u'w'} = \rho K_M \frac{d\bar{u}}{dz}; \tag{1-15}$$

then K_M is called the *eddy viscosity coefficient*, which has the same dimension as the kinematic coefficient of viscosity ν.

In general, since $K_M \gg \nu$, we may use this eddy viscosity term while omitting the molecular viscosity term. Then Eq. (1-13) and the Navier–Stokes equation, which includes Eq. (1-13) implicitly, will have the same form as Eq. (1-11) if the velocity v is replaced by the mean velocity $\bar{v}$.

As is known from Eq. (1-14), the Reynolds stress is a tensor, so that the eddy viscosity coefficient should also be a tensor: the vertical and horizontal coefficients are different in their orders of magnitude. It is to be noted that the horizontal coefficient varies greatly depending on the scale of the motion, whereas thermal stratification plays an important role in determining the order of magnitude of the vertical coefficient.

If we take the concentration of a substance for G, Eq. (1-2) will result in the *diffusion equation*. In this case the eddy viscosity coefficient will be replaced by the *eddy diffusion coefficient*, or the *eddy diffusivity*. Since the concentration of a substance is a scalar quantity while the momentum is a vector quantity, the eddy diffusivity is a vector quantity.

1.2.5 Nondimensionalized equations and nondimensional quantities

Any physical quantity has dimension(s) and unit(s). There are three mutually

---

[1]Reynolds' axiom:

$$\bar{\bar{a}} = \bar{a}, \quad \overline{a+b} = \bar{a}+\bar{b}, \quad \overline{\bar{a}b} = \bar{a}\bar{b}, \quad \frac{\overline{\partial a}}{\partial s} = \frac{\partial\bar{a}}{\partial s}, \quad \overline{\int_\alpha^\beta a\,ds} = \int_\alpha^\beta \bar{a}\,ds. \tag{1-A}$$

independent dimensions. Usually we adopt the length [L], the time [T] and the mass [M] as fundamental dimensions. Other dimensions, for instance those for temperature or electricity, can be constructed by a combination of these three, when necessary.

Nondimensionalization of a system of equations provides the following advantages.

(1) The equations become independent of units by which the variables are measured.

(2) If a certain number of dimensions are eliminated, the number of equations is also reduced by the same number, which makes the system of treatment much simpler.

(3) It is easier to compare the relative size or importance of terms in the equations, to examine thus selected significant terms.

(4) It is useful to derive semi-empirical formulas from the data of a complex phenomenon like turbulent motion, which are very difficult to be obtained analytically.

A simple example of nondimensionalization is given below. First, we consider the N–S equation for an incompressible fluid, in which we let p include the gravity term $\rho g z$ with g the acceleration of gravity, in addition to the pure pressure p:

$$\frac{\partial v}{\partial t} + (v \bullet \nabla)v = -\frac{1}{\rho}\nabla p + \nu \nabla^2 v \qquad (1\text{-}16)$$

and the equation of continuity for incompressible fluid

$$\nabla \bullet v = 0 . \qquad (1\text{-}17)$$

If we use nondimensional variables,

$$v^* = \frac{v}{U}, \quad t^* = \frac{U}{L}t, \quad x^* = \frac{x}{L}, \quad p^* = \frac{p - p_0}{\rho U^2} \qquad (1\text{-}18)$$

where L, U and p_0 are representative scales of length, velocity and a standard pressure, respectively, then Eqs. (1-16) and (1-17) become

$$\frac{\partial v^*}{\partial t^*} + \left(v^* \bullet \nabla^*\right)v^* = -\nabla^* p^* + \frac{1}{Re}\nabla^{*2}v^* , \qquad (1\text{-}19)$$

$$\nabla^* \bullet v^* = 0 \qquad (1\text{-}20)$$

where ∇^* is the gradient operator with respect to x^*, and

$$Re = \frac{\rho U L}{\mu} = \frac{U L}{v} \qquad (1\text{-}21)$$

is the *Reynolds number*, *Re*, which has no dimension. For the same value of *Re*, the equation will have the same solution, even if *L*, *U* and/or *v* has different values. This is the *Reynolds' similarity law*, which provides the basis for model experiments.

The second term on the right-hand side of Eq. (1-19) has a form of the inertia term divided by the viscous term in Eq. (1-16), so that the Reynolds number is the representative ratio of the two terms. Therefore, if *Re* is very small compared with unity, we can discuss the equation by omitting the inertia term.

It was relatively simple to nondimensionalize Eq. (1-16). However, if we want to do the same thing for the case of a stratified fluid on the rotating earth, or for the *geophysical fluid*, the derivation will be much more complex. The equation of motion on the local Cartesian coordinate with the Coriolis acceleration term, the gravity term and the eddy viscosity term is, under the *Boussinesq approximation*[2]

$$\frac{\partial v}{\partial t} + (v \bullet \nabla)v + f_0 k \times v = -\frac{1}{\rho_0}\nabla p' - \frac{\rho'}{\rho_0}gk + K_M \nabla^2 v , \qquad (1\text{-}22)$$

where ρ_0 is the representative density, f_0 the Coriolis parameter f at the representative latitude φ_0, k the vertically upward unit vector, and p' and ρ' are the perturbations of p and ρ from the reference sea in which the hydrostatic equilibrium state with the following balance holds:

$$\frac{dp_r}{dz} = -\rho_r g. \qquad (1\text{-}23)$$

Let us further use the representative length *L*, time *T* (e.g., the period for periodical motion), velocity *U*, and the Brunt–Väisälä frequency *N*, to derive the nondimensional variables

$$\left.\begin{array}{ll} t^* = \dfrac{t}{T}, & x^* = \dfrac{x}{L}, \quad v^* = \dfrac{v}{U} \\[2mm] p'^* = \dfrac{p'}{\rho_0 f_0 U L}, & g^* = \dfrac{g}{NU}, \quad \rho'^* = \dfrac{\rho'}{\rho_0} \end{array}\right\} . \qquad (1\text{-}24)$$

---

[2] The approximation in which the variation of density is taken into account only for the buoyancy related term including the gravity g; elsewhere the representative density ρ_0, which is a constant, is used.

Then rewriting Eq. (1-22) leads to:

$$\varepsilon_T \frac{\partial v^*}{\partial t^*} + \varepsilon\left(v^* \bullet \nabla^*\right)v^* + k \times v^* = -\nabla^* p'^* + Bp'^* g^* k + E\nabla^{*2} v^* \qquad (1\text{-}25)$$

where the nondimensional numbers appearing in the equation are:

$$\varepsilon = \frac{U}{f_0 L} \qquad : \text{Rossby number}, \qquad\qquad\qquad (1\text{-}26)$$

$$E = \frac{K_M}{f_0 L^2} \qquad : \text{Ekman number}, \qquad\qquad\qquad (1\text{-}27)$$

$$\varepsilon_T = \frac{1}{f_0 T}, \qquad\qquad\qquad (1\text{-}28)$$

$$B = \frac{N}{f}, \qquad\qquad\qquad (1\text{-}29)$$

$$g^* = \frac{g}{NU}. \qquad\qquad\qquad (1\text{-}30)$$

Equation (1-25) corresponds to the form in which each term in the original Eq. (1-22) has been normalized by the Coriolis term. Once the variables are nondimensionalized, the asterisk that represents nondimensionality is customarily omitted.

In the large scale oceanic or atmospheric motions on the earth, the Coriolis term balances with the pressure gradient term; this condition is what we call the *geostrophic balance*. In this way, the normalization by using the Coriolis term will make it easier to discuss deviations from the geostrophic balance. The Rossby number ε and the Ekman number E represent the importance of the inertia and viscous terms, respectively, relative to the Coriolis term.

The *Brunt–Väisälä frequency* N is defined by

$$N = \left(-\frac{g}{\rho_0}\frac{\partial \overline{\rho}}{\partial z} - \frac{g^2}{c^2}\right)^{1/2} \qquad\qquad\qquad (1\text{-}31)$$

where $\overline{\rho}$ is the mean value of ρ, which may vary horizontally or with time, and c is the sound speed. The term g^2/c^2 corresponds to the density gradient when a parcel of fluid moves adiabatically in the vertical direction, and it can be ignored where the stratification is strong. N is the frequency of oscillation caused by the restoring force of gravity when a column of fluid is displaced infinitesimally in the vertical direction in a stratified fluid. B in Eq. (1-29) represents the ratio of this frequency by stratification to the *inertial frequency* caused by the Coriolis force. When B is small, the effect of stratification can be ignored.

The boundary layers are domains where the term E becomes significant. Processes of variation proceed by conditions where the term of ε is significant. In the range where these nondimensional numbers are not large, we can analyze phenomena by perturbations of terms including these numbers. Examples of such analyses will be described and discussed in the following chapters in detail.

REFERENCES

Gill, A. E. (1981): *Atmosphere–Ocean Dynamics*, Academic Press, 662 pp.

Oceanographic Society of Japan (ed.) (1991): *The Oceans and Global Environment—Frontier of Oceanography*, Univ. of Tokyo Press, Tokyo, 149 pp. (in Japanese).

Sverdrup, H. U. *et al.* (1942): *The Oceans—Their Physics, Chemistry and General Biology*, Prentice-Hall, 1087 pp.

Toba, Y. (1970): Air-sea boundary processes. In *Kaiyo Kagaku Kiso Koza (Fundamental Course in Oceanographic Sciences) 1, Physical Oceanography I*, ed. by J. Masusawa, Tokai University Press, Tokyo, Japan, pp. 145–263. (in Japanese).

Chapter 2

Air–Sea Interface Processes and Wind Waves

Yoshiaki TOBA

Ocean–Atmosphere Interactions, Ed. Y. Toba, pp. 13–62.
© by TERRAPUB / Kluwer, 2003.

Air–Sea Interface Processes and Wind Waves

Yoshiaki TOBA

> *In a season of calm weather*
> *Though inland far we be,*
> *Our souls have sight of that immortal sea*
> *Which brought us hither,*
> *Can in a moment travel thither,*
> *... And hear the mighty waters rolling evermore.*
>
> —*William Wordsworth*

2.1. METHOD OF NONDIMENSIONALIZATION

The usefulness of nondimensionalizing equations was described in the previous chapter. The method of nondimensionalization is especially useful for processes of the boundary layers where turbulence dominates. In this section, general issues of nondimensionalizatiton will be reviewed.

2.1.1 Laws on reduction of the number of dimensions

The *Pi theorem* is concerned with reduction of the number of dimensions. Suppose we have n physical quantities Q_j ($j \leq n$), and r fundamental dimensions q_i ($i \leq r$). The content of the Pi theorem is expressed as follows. If there is a functional relation such that

$$f\left(Q_1, Q_2, \cdots, Q_n\right) = 0, \tag{2-1}$$

then this is equivalent to

$$F\left(\Pi_1, \Pi_2, \cdots, \Pi_{n-r}\right) = 0 \tag{2-2}$$

where the Π's are the $n - r$ independent nondimensional products, which were constructed by combining quantities not more than $r + 1$ among Q_j.

This can be understood by the following interpretation. Since Q_j is written as a product of fundamental dimensions q_i:

$$
\left.
\begin{aligned}
Q_1 &= A q_1^{\alpha_1} q_2^{\alpha_2} \cdots q_r^{\alpha_r} \\
Q_2 &= B q_1^{\beta_1} q_2^{\beta_2} \cdots q_r^{\beta_r} \\
&\quad\cdots \\
&\quad\cdots \\
Q_n &= N q_1^{\nu_1} q_2^{\nu_2} \cdots q_r^{\nu_r}
\end{aligned}
\right\}
\qquad (2\text{-}3)
$$

with A, B, $\cdots$, N being numerical values, we can eliminate r fundamental dimensions out of any $r + 1$ combinations of Q_j, to obtain one nondimensional product. This procedure can be successively repeated until $n - r$ mutually independent nondimensional products are obtained. Namely, the nondimensional products thus obtained can be written as

$$
\left.
\begin{aligned}
\Pi_1 &= \Pi_1(Q_1,\, Q_2,\, \cdots,\, Q_r;\, Q_{r+1}) \\
\Pi_2 &= \Pi_2(Q_1, Q_2,\, \cdots,\, Q_r;\, Q_{r+2}) \\
&\quad\cdots \\
&\quad\cdots \\
\Pi_{n-r} &= \Pi_{n-r}(Q_1, Q_2,\, \cdots,\, Q_r;\, Q_n)
\end{aligned}
\right\}.
\qquad (2\text{-}4)
$$

In this case, Q_1, Q_2, $\cdots$, Q_r, are common. However, since Q_{r+1}, Q_{r+2}, $\cdots$, Q_n are independent, the products Π_1, Π_2, $\cdots$, Π_{n-1} are independent. If we consider another Π, it can be derived from the above independent relations, for instance, as $\Pi(Q_1, Q_2, \cdots, Q_{r-1}, Q_{r+1}, Q_{r+2})$ by eliminating Q_r from Π_1 and Π_2. Therefore, Eqs. (2-1) and (2-2) are equivalent.

It is usual to take the length L, the time T and the mass M as fundamental dimensions q_i. Attention should be paid to the fact that even though the number of q_i appears to be r, the real number of fundamental quantities could be $r - 1$, if one of them is dependent on another. One example of such a case is when T appears only in the combination of LT^{-2}.

However, the Pi theorem does not tell how to combine the variables to make nondimensional products or how to select mutually independent nondimensional variables. In order to make meaningful nondimensional variables, we must invoke our physical insight. How this can be done will be discussed in the following section.

2.1.2 Procedures for making nondimensional variables

First we classify the variables relevant to the phenomenon under consideration into the followings, taking their physical meaning into account. (See, e.g., Eq. (2-6).)

a: *Independent variables*—such as time and space.

b: *Dependent variables*—unknowns.

c: *External conditions*—variables implying factors that control the phenomenon.

d: Parameters like *physical constants*—some variables become constant if the phenomenon is concerned with a specific material such as sea water, or conditions such as the motion occurring only on the earth. The same parameters could be classified as c, according to the case.

e: *Parameters representing scales*—time and space scales representative of the phenomenon, or scales to become standard units of time and space for the phenomenon.

Next, the following three steps will be carried out.

(1) Simplification of variables

Variables are simplified as far as possible. For instance, we may divide μ and p by ρ so that ρ will not appear explicitly for, say, a kinematic treatment.

(2) Nondimensionalization of independent and dependent variables

First, a and b are to be nondimensionalized by use of e. If that is not possible, d and e, or c, d and e should be used for the nondimensionalization. We have now independent and dependent variables without dimensions: Π_1, Π_2, ...

(3) Nondimensionalization of external conditions

The variables in c are to be nondimensionalized using d and e. The resulting nondimensional parameters control the conditions of the motion under consideration. They are called *conditioning (environmental) parameters*.

Let us take, as an example, the velocity distribution of the flow in a circular tube to perform the above procedures. What we should derive is the form of the flow velocity u, which is an unknown dependent variable, as a function of the distance from the center of the tube r, which is an independent variable. The factors that determine the conditions are the horizontal pressure gradient G ($\equiv \partial p/\partial x$), the density of the material ρ, the coefficient of viscosity μ, and the radius of the tube a. First, consider a stationary laminar flow. Since the acceleration is zero, the equation of motion is just a balance between the forces without the acceleration term. Therefore the density ρ here is irrelevant. Eq. (2-1) in this case will be

$$f(r, u, G, \mu, a) = 0, \qquad\qquad (2 - 5)$$

and the procedure can be written as

$$\begin{bmatrix} a \\ b \\ c \\ d \\ e \end{bmatrix} = \begin{bmatrix} r \\ u \\ G \\ \mu \\ a \end{bmatrix} \qquad (2\text{-}6)$$

where $n = 5$ and $r = 3$, so that the number of nondimensional variables is $n - r = 2$. Following procedure (2), we obtain

$$\Pi_1 = \frac{r}{a} \equiv r^* \quad \text{(Independent variables)}. \qquad (2\text{-}7)$$

Suppose u is of the first order; then

$$\Pi_2 = a^x \mu^y G^z u$$

and the equation for the dimension will be

$$[L]^x [ML^{-1}T^{-1}]^y [ML^{-2}T^{-2}]^z [LT^{-1}] = [LMT]^0.$$

From this, we can derive $x = -2$, $y = 1$, $z = -1$, and

$$\Pi_2 = \frac{\mu u}{a^2 G} \equiv u^* \quad \text{(Dependent variables)}. \qquad (2\text{-}8)$$

That is, Eq. (2-2) becomes

$$f(u^*, r^*) = 0 \quad \text{or} \quad u^* = f(r^*) \qquad (2\text{-}9)$$

or in dimensional variables

$$\frac{\mu u}{a^2 G} = f\left(\frac{r}{a}\right). \qquad (2\text{-}10)$$

The functional form of f could be analytically solved since a laminar flow is considered here, as

$$-\frac{\mu u}{a^2 G} = \frac{1}{4}\left\{1 - \left(\frac{r}{a}\right)^2\right\} \qquad (2\text{-}11)$$

or

$$u = -\frac{G}{4\mu}\left(a^2 - r^2\right).$$

(2-12)

This is called the *Poiseuille law*.

The dimensional analysis is not really necessary for a laminar flow, which has an analytic solution. It will be more useful for a turbulent flow that cannot be handled analytically.

If u^* is obtained experimentally and plotted against r^*, its values will predictably fall into the curve expressed by Eq. (2-12) if the flow is laminar, but such agreement cannot be expected when the flow is turbulent. A turbulent flow has acceleration in it, and ρ should be added to the variables. Eq. (2-6) will become

$$\begin{bmatrix} a \\ b \\ c \\ d \\ e \end{bmatrix} = \begin{bmatrix} r \\ u \\ G \\ \rho, \mu \\ a \end{bmatrix}.$$

(2-13)

We now need Π_3. Since this must be a conditioning parameter, c should be nondimensionalized by d and e following procedure (3). We find that

$$\Pi_3 = \frac{\rho a^3 G}{\mu^2}.$$

(2-14)

For a laminar flow where the current velocity at $r = 0$ is u_0, we obtain from Eq. (2-12)

$$u_0 = -\frac{1}{4}\frac{a^2 G}{\mu},$$

(2-15)

so that Π_3 is proportional to au_0/v and it is essentially nothing but the Reynolds number. That is,

$$u^* = f\left(r^*, Re\right), \quad Re = \frac{\rho a^3 G}{\mu^2}.$$

(2-16)

Thus, experimental results will supposedly have systematic deviation from the laminar case if Π_3 is selected as a varying parameter, and they would be better classified as such. The framework of Eq. (2-16) will be given first, and the specific form of f will be determined empirically afterwards. In this sense the final form of f is called a *semi-empirical* one.

As seen above, even when experimental data in dimensional quantities are so complex that it is almost impossible to deduce any law from them, it becomes easy to extract an essential law by reducing variables in dimensionless forms.

2.2 MACROSCOPIC STRUCTURE OF THE BOUNDARY LAYER

Let us apply the techniques discussed in the previous section to the derivation of the macroscopic structure of the boundary layer. The sea surface is generally covered by waves, which include *wind waves*. Wind waves are special hydrodynamic phenomena that connect two boundary layers of the air and the water. This will be discussed in Section 2.4, after the dynamics of water waves are reviewed. To begin with, in this section the general structure of the boundary layer on a solid surface will be considered.

2.2.1 Structure of flow on a solid surface

Since the density of water is very much larger than that of air, we will first consider a flow on a solid surface. Suppose the radius of the tube discussed in Section 2.1.2 is very large, and consider the structure of the mean flow u near the boundary $z = 0$. Since the average state is stationary, and $\bar{u} = \bar{u}(z)$, $\bar{v} = \bar{w} = 0$, the vertical flux in this region of the momentum in the x-direction is equal to the wind stress τ, or the absorption rate of the air momentum, at the boundary. That is, from Eqs. (1-13) and (1-14), we obtain

$$\mu \frac{d\bar{u}}{dz} - \rho \overline{u'w'} = \tau. \qquad (2\text{-}17)$$

In view of the structure of the flow, the variables are classified as

$$\begin{bmatrix} a \\ b \\ c \\ d \\ e \end{bmatrix} = \begin{bmatrix} z \\ \bar{u} \\ \tau \\ \rho, \mu \\ — \end{bmatrix}. \qquad (2\text{-}18)$$

First, following procedure (1), we convert this into

$$\begin{bmatrix} a \\ b \\ c \\ d \\ e \end{bmatrix} = \begin{bmatrix} z \\ \overline{u} \\ u_* \\ v \\ - \end{bmatrix} \qquad (2\text{-}19)$$

where τ is replaced by the *friction velocity*, defined by

$$u_* = \sqrt{\frac{\tau}{\rho}}. \qquad (2\text{-}20)$$

This u_* will be used throughout this chapter as the most important conditioning (environmental) parameter to represent the boundary layer condition for a quasi-stationary boundary layer flow.

The nondimensional variables are determined by procedure (2) after taking into account that $n - r = 2$;

Independent variables: $\Pi_1 = \Pi_1(v, u_*; z) = \dfrac{u_* z}{v}$,

Dependent variables: $\Pi_2 = \Pi_2(u_*; \overline{u}) = \dfrac{\overline{u}}{u_*}$.

Therefore, we obtain

$$\frac{\overline{u}(z)}{u_*} = f\left(\frac{u_* z}{v}\right). \qquad (2\text{-}21)$$

This formula corresponds to the *law of the wall* originally given by Prandtl (1932).

Equation (2-21) holds when the boundary is accurately expressed as $z = 0$, or it is a smooth surface, but if it is a rough surface, another variable to represent the roughness should be added, and Π_3 becomes necessary as will be discussed later.

For a *viscous sublayer*, where Eq. (2-17) can be written as

$$\mu \frac{d\overline{u}}{dz} = \tau,$$

we can assume $\bar{u} = 0$ at $z = 0$, and an analytical solution can be obtained,

$$\frac{\bar{u}}{u_*} = \frac{u_* z}{v} . \qquad (2\text{-}22)$$

If we consider the thickness of the sublayer δ_v,

$$\begin{bmatrix} a \\ b \\ c \\ d \\ e \end{bmatrix} = \begin{bmatrix} - \\ \delta_v \\ u_* \\ v \\ - \end{bmatrix} \qquad (2\text{-}23)$$

which leads to $n - r = 1$, and we get

$$\Pi_1 = \frac{u_* \delta_v}{v} = \alpha_v . \qquad (2\text{-}24)$$

Since Π_1 is not a function of any variable, α_v should be a universal constant. α_v is semi-empirically known to be of the order of 5, and we obtain

$$\delta_v = \frac{5v}{u_*} . \qquad (2\text{-}25)$$

Next, consider the *turbulent region* where z is large and the situation is written as

$$\tau = -\rho \overline{u' w'} \quad \text{or} \quad u_*^2 = -\overline{u' w'} . \qquad (2\text{-}26)$$

Here since the boundary condition, such as $\bar{u} = 0$ at $z = 0$, cannot be determined, we take the derivative of $\bar{u}$ as an unknown variable, and write the system of the variables as

$$\begin{bmatrix} a \\ b \\ c \\ d \\ e \end{bmatrix} = \begin{bmatrix} z \\ \dfrac{d\overline{u}}{dz} \\ u_* \\ - \\ - \end{bmatrix}. \tag{2-27}$$

Then we obtain

$$\Pi_1 = \frac{d\overline{u}}{dz} \bigg/ \frac{u_*}{z} = \frac{1}{k}. \tag{2-28}$$

Now $1/k$ is a universal constant. k is the *von Kármán constant* ($=0.4$). If we integrate Eq. (2-28) and consult Eq. (2-21), we obtain the *logarithmic boundary layer* relationship

$$\frac{\overline{u}(z)}{u_*} = \frac{1}{k} \ln \frac{u_* z}{v} + C = \frac{1}{k} \ln \frac{u_* z}{\beta v}. \tag{2-29}$$

The integral constant C is semi-empirically known to be 5.5 ($\beta = 0.111$). The lower limit for the turbulent boundary layer δ_1 can be determined in the same manner as δ_v (see Fig. 2-1), as

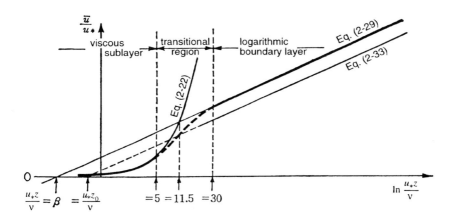

Fig. 2-1. Velocity profile in the boundary layer.

$$\delta_1 = \alpha_1 \frac{v}{u_*}, \quad \alpha_1 = 30. \tag{2-30}$$

From the definition (1-15) of the *eddy viscosity coefficient* K_M, together with Eq. (2-28), the form of K_M in the boundary layer can be derived as

$$K_M = ku_*z. \tag{2-31}$$

If the solid surface is uneven, a variable expressing the roughness is added. If we take as the variable the average height ε of the roughness, Eq. (2-19) will be replaced by

$$\begin{bmatrix} a \\ b \\ c \\ d \\ e \end{bmatrix} = \begin{bmatrix} z \\ \bar{u} \\ u_*, \varepsilon \\ v \\ - \end{bmatrix}$$

and

$$\text{Conditioning parameter:} \quad \Pi_3 = \Pi_3(v, u_*; \varepsilon) = \frac{u_* \varepsilon}{v}$$

will be added by procedure (3). Now we have

$$\frac{\bar{u}(z)}{u_*} = f\left(\frac{u_* z}{v}, \frac{u_* \varepsilon}{v}\right). \tag{2-32}$$

The parameter $u_* \varepsilon / v$ is called the *roughness Reynolds number*.

In the turbulent region, Eq. (2-28) should be correct, and C in Eq. (2-29) should vary according to Π_3 so that both equations hold. That is, as seen in Fig. 2-1, the logarithmic profile shifts to the lower side in a parallel manner with $u_* \varepsilon / v$. When $u_* \varepsilon / v$ is smaller than approximately 4, the effect of the roughness disappears. This part of the logarithmic law, after the parallel downward shift, is customarily written as

$$\frac{\bar{u}}{u_*} = \frac{1}{k} \ln \frac{z}{z_0}, \tag{2-33}$$

where $\bar{u} = 0$ is assumed at $z = z_0$. z_0 is called the *roughness parameter* or the *roughness length.* The nondimensional form of z_0 is u_*z_0/v, which represents the size of the parallel downward shift. Since this should be determined by the conditioning parameter $u_*\varepsilon/v$, there must be a universal relationship between u_*z_0/v and $u_*\varepsilon/v$. This reasoning may be justified by a relationship that is empirically accepted,

$$\frac{z_0}{\varepsilon} = \frac{1}{30}. \tag{2-34}$$

The above discussion applies to the natural roughness. More generally, the situation depends on the type of roughness.

2.2.2 Thermal stratification case

If thermal stratification exists, besides the vertical momentum transport τ, the vertical heat transport H, the acceleration of gravity g, the specific heat at constant pressure C_p, and the absolute temperature T are to be added. The system of variables will be

$$\begin{bmatrix} a \\ b \\ c \\ d \\ e \end{bmatrix} = \begin{bmatrix} z \\ \bar{u}, \bar{T}, \bar{\rho} \\ \tau, H \\ \mu, g, C_p \\ - \end{bmatrix}. \tag{2-35}$$

If we consider the unknown $\bar{u}$, we move $\bar{T}$ and $\bar{\rho}$ to d with their representative values T_0 and ρ_0. Reclassifying the variables, an additional nondimensional number Π_2 will be added to Eq. (2-28). This number is the ratio of the work done by the buoyancy force per unit mass $gH/\rho_0 C_p T_0$ in the turbulence field to the work done by the Reynolds stress per unit mass (neutral stratification is assumed):

$$u_*^2 \frac{d\bar{u}}{dz} = \frac{u_*^3}{kz}. \tag{2-35'}$$

That is,

$$\Pi_2 = \zeta = -\frac{(g/T_0)(H/\rho_0 C_p)}{u_*^3/kz}. \tag{2-36}$$

ζ is called the *stability function,* being essentially the same as the *Richardson number Ri.*

For the profile of $\bar{u}$ in the turbulence region, Π_1 in Eq. (2-28) is no longer a universal constant but a function of Π_2, and we have

$$\frac{d\bar{u}}{dz} \bigg/ \frac{u_*}{z} = f(\zeta). \tag{2-37}$$

That is,

$$\frac{d\bar{u}}{dz} = \frac{u_*}{kz} \phi(\zeta). \tag{2-38}$$

As a matter of fact, the functional form of $\phi(\zeta)$ is given by the solution of an equation that was derived through a theory containing physical hypotheses such as the mixing length,

$$\phi^4 - \sigma\zeta\phi^3 - 1 = 0 \tag{2-39}$$

where σ is a numerical coefficient determined by observations. Eq. (2-39) is called the KEYPS equation. If ζ is near zero, or the stability is almost neutral, we can use the first order Taylor expansion of the $\phi(\zeta)$ around $\zeta = 0$. This leads to the *log-linear law*

$$\frac{\bar{u}}{u_*} = \frac{1}{k} \left(\ln \frac{z}{z_0} + \beta \frac{z}{L} \right), \tag{2-40}$$

where $L = z/\zeta$ is known as the *stability length* (the dynamical thickness of turbulent layer) proposed by Monin and Obukhov (1954) and β is a semi-empirical coefficient to be determined by observations.

2.3 DYNAMICS OF WATER SURFACE WAVES

When undulations are caused on the water surface, they propagate due to the restoring forces of gravity and surface tension. Before we discuss the generation of wind waves and the particular flow structure associated with them, we are going to summarize in this section the fundamental issues of the simple water surface waves.

2.3.1 System of equations for water surface waves

We consider surface waves which can be generated on incompressible and inviscid water. Then, since the motion is irrotational, we can define the *velocity potential* ϕ by $\boldsymbol{u} = \nabla\phi$, which, together with the incompressibility condition $\nabla \cdot \boldsymbol{u} = 0$, will lead to the Laplace equation

$$\nabla^2 \phi = 0. \tag{2-41}$$

If we take x and y axes on the still water surface, and a z axis to the vertical upward, then the boundary condition at the bottom ($z = -h$) will be

$$\frac{\partial \phi}{\partial z} = w = 0. \tag{2-42}$$

If we assume that water particles on the surface will always remain on the surface (although this condition fails to be satisfied when water waves break at the surface), and the water surface is expressed by $z = \zeta(x, y, t)$, then

$$\left[\frac{\partial \phi}{\partial z}\right]_\zeta = \frac{\partial \zeta}{\partial t} + [\nabla_h \phi]_\zeta \cdot [\nabla_h \zeta], \tag{2-43}$$

where $\nabla_h = (\partial/\partial x, \partial/\partial y)$. Since Eq. (2-43) contains an unknown ζ, another condition is necessary. For this purpose, one of the following two equations will be used. Applying the pressure equation to $z = \zeta$ will provide

$$\frac{p}{\rho_w} + g\zeta + \left[\frac{\partial \phi}{\partial t}\right]_\zeta + \frac{1}{2}[\nabla \phi]_\zeta^2 = 0, \tag{2-44}$$

while applying the Lagrangian derivative of the pressure equation to $z = \zeta$ will result in (Phillips, 1977)

$$\left[\frac{D}{Dt}\left(\frac{p}{\rho_w}\right)\right]_\zeta + \left[\frac{\partial^2 \phi}{\partial t^2} + g\frac{\partial \phi}{\partial z}\right]_\zeta + \left[\frac{\partial u^2}{\partial t}\right]_\zeta + \frac{1}{2}[u \cdot \nabla u^2]_\zeta = 0 \tag{2-45}$$

where ρ_w is the density of water. If we consider the surface tension S, the pressure becomes

$$p = p_a - S\left(\frac{1}{R_1} + \frac{1}{R_2}\right) = p_a - SF(\zeta), \quad F(\zeta) = \frac{\nabla_h^2 \zeta}{\left\{1 + (\nabla_h \zeta)^2\right\}^{3/2}} \tag{2-46}$$

where R_1 and R_2 are the radii of curvature and p_a is the atmospheric pressure.

Attention should be paid to the fact that Eqs. (2-43), (2-44) and (2-45) are nonlinear, and the above boundary condition is given at the unknown free surface $z = \zeta(x, y, t)$. Therefore, we try to solve the equations using the method of successive approximation by first obtaining the linear solution. If we expand Eq.

(2-45) around $z = 0$ as

$$\frac{D}{Dt}\left(\frac{p_a}{\rho_w}\right) + \left[\left(1 + \zeta\frac{\partial}{\partial z} + \frac{1}{2}\zeta^2\frac{\partial^2}{\partial z^2} + \cdots\right)\left(\frac{\partial^2\phi}{\partial t^2} + g\frac{\partial\phi}{\partial z} - \frac{S}{\rho_w}\frac{D}{Dt}F(\zeta)\right)\right]_0$$

$$+\left[\left(1 + \zeta\frac{\partial}{\partial z} + \cdots\right)\frac{\partial u^2}{\partial t}\right]_0 + \left[\frac{1}{2}\boldsymbol{u}\bullet\nabla u^2 + \cdots\right]_0 = 0 \qquad (2\text{-}47)$$

and choose a parameter $\varepsilon = ak$, where a is the wave amplitude and k is the absolute value of the wave number, then expanding ϕ, $\boldsymbol{u}$, ζ as, e.g.,

$$\phi = \varepsilon\phi_1 + \varepsilon^2\phi_2 + \varepsilon^3\phi_3 + \cdots, \qquad (2\text{-}48)$$

we obtain the following set of linear equations, after we retain the terms proportional to ε only. It should be noted that ε corresponds to the *wave steepness* for waves of infinitesimal amplitude (or linear waves).

$$\boldsymbol{u}_1 = \nabla\phi_1, \quad \nabla^2\phi_1 = 0, \qquad (2\text{-}49)$$

$$\frac{\partial\phi_1}{\partial z} = 0, \quad z = -h, \qquad (2\text{-}50)$$

$$\frac{\partial\zeta_1}{\partial t} = \frac{\partial\phi_1}{\partial z}, \quad z = 0, \qquad (2\text{-}51)$$

$$\frac{\partial^2\phi_1}{\partial t^2} + g\frac{\partial\phi_1}{\partial z} - \frac{S}{\rho_w}\nabla_h^2\left(\frac{\partial\phi_1}{\partial z}\right) = 0, \quad z = 0. \qquad (2\text{-}52)$$

The solution of these equations will give the *infinitesimal wave* (*linear wave*). Stokes (1847) first suggested the method of deriving higher order solutions of a finite amplitude wave, by using the linear solution as a first approximation, and so this wave is sometimes called the *Stokes wave*.

2.3.2 Infinitesimal waves and finite amplitude waves

The first approximation solution of the infinitesimal waves is assumed for the surface displacement ζ to be

$$\zeta = a\cos(\boldsymbol{k}\bullet\boldsymbol{x} - \sigma t). \qquad (2\text{-}53)$$

Then the corresponding velocity potential is

$$\phi = \frac{\sigma a \cosh k(z+h)}{k \sinh kh} \sin(k \cdot x - \sigma t), \qquad (2\text{-}54)$$

where k is the wave number vector, $|k| = k = 2\pi/\lambda$, λ being the wavelength, σ the angular frequency ($=2\pi/T$), and T the period. The *dispersion relationship* is

$$\sigma^2(k) = gk \tanh kh. \qquad (2\text{-}55)$$

The *phase speed* is

$$c = \frac{\sigma}{k} = \sqrt{\frac{g}{k} \tanh kh} \ . \qquad (2\text{-}56)$$

The effect of the surface tension does not appear in Eq. (2-55), but if we replace g by

$$g_* = g + \frac{Sk^2}{\rho_w} \qquad (2\text{-}57)$$

it can be included.

If the water depth h is large compared with the wavelength ($kh \gg 1$), Eqs. (2-54), (2-55) and (2-56) will have simplified forms of *deep-water waves* as follows:

$$\phi = k^{-1}\sigma a e^{kz} \sin(k \cdot x - \sigma t), \qquad (2\text{-}58)$$

$$\sigma^2 = gk, \quad c = |c| = \frac{\sigma}{k} = \frac{g}{\sigma} = \sqrt{\frac{g}{k}} = \sqrt{\frac{g\lambda}{2\pi}} \ . \qquad (2\text{-}59)$$

Subsequent discussions will mainly be on deep-water waves.

When the effect of surface tension and the effect of gravity are of the same order, or at the wavelength $\lambda = 2\pi(S/\rho_w g)^{1/2} = \lambda_m$, the phase speed will have a minimum value c_m. For water of 20°C, $\sigma_m = 84.8$ rad s^{-1}, $\lambda_m = 1.71$ cm, and $c_m = 23.1$ cm s^{-1}.

The irrotational, *finite amplitude* gravity waves have the third order free surface form

$$\zeta = a \cos kx + \frac{1}{2}a^2 k \cos 2kx + \frac{3}{8}a^3 k^2 \cos 3kx + O(a^4 k^3). \qquad (2\text{-}60)$$

This wave profile is identical, to the third order, with a trochoidal curve, the crest sharpened and the trough flattened. The phase speed is given by

$$
\left.
\begin{aligned}
c^2 &= \frac{g}{k}\left\{1 + a^2k^2 + O\left(a^4k^4\right)\right\} \simeq \frac{g}{k}\left(1 + \pi^2\delta^2\right) \\[2ex]
\delta &= \frac{2a}{\lambda} = \frac{ak}{\pi}
\end{aligned}
\right\}
\tag{2-61}
$$

where δ is the *wave steepness*. Thus, the wave speed depends on the wave amplitude, and it is slightly greater than that of the linear wave.

These waves have a maximum wave height (or wave steepness) when $\delta = 0.142 (\approx 1/7)$. This is *Stokes' limiting criterion for wave breaking*. At this condition, the crest is sharpened to have an included angle of 120 degrees, while the particle speed on the surface becomes the same as that of the phase speed ($c = 1.10(g/k)^{1/2}$) and the acceleration at the crest is $g/2$ in all directions within the angle of 120 degrees. This maximum wave profile can be given to a very high accuracy by the form (Longuet-Higgins, 1972)

$$
\frac{d\zeta}{dx} = \tan\left(\frac{\pi x}{3\lambda}\right), \quad |x| \le \frac{\lambda}{2}.
\tag{2-62}
$$

Schwartz (1974) derived these waves in another higher order expansion form, which can be applied to the finite depth case as well. Figure 2-2 shows the calculated results.

If the surface *wind drift* is present on the water surface, waves may break at less steep conditions. Banner and Phillips (1974) derived the following relationship, using the condition of the surface particle and the phase speed being equal, on the assumption that a layer of the surface wind drift superposes on irrotational waves:

$$
\zeta_{max} = \frac{c^2}{2g}\left(1 - \frac{q_0}{c}\right)^2
\tag{2-63}
$$

where ζ_{max} is the maximum water level and q_0 the magnitude of the surface drift at the mean water level. Wind wave breaking in nature is not so simple, yet it is actually observed that wave breaking occurs at a wave height much smaller than Stokes' limiting wave height.

Brooke Benjamin and Feir (1967) found out that a train of one-dimensional finite amplitude waves is unstable for a small perturbation of wave amplitude. This is called *Benjamin–Feir instability* or *side-band instability* of the Stokes wave. Later a two-dimensional wave train was found by McLean (1982) to be unstable theoretically. These results have been proved experimentally, too. The

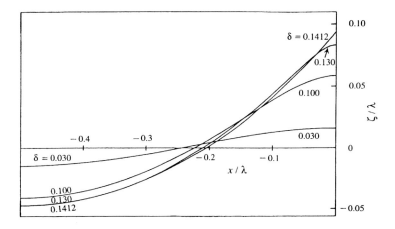

Fig. 2-2. Wave form of Stokes' expansion for gravity waves (after Schwartz, 1974). Vertical and horizontal axes are normalized by the wave length.

instability of finite amplitude waves will develop to cause wave breaking.

A finite amplitude capillary wave is known to have an exact analytical solution (Crapper, 1957). The wave profile ζ is given as a function of x through a parameter α as follows:

$$\left. \begin{aligned} \frac{x}{\lambda} &= \alpha - \frac{2}{\pi} \frac{A \sin 2\pi\alpha}{1 + A^2 + 2A \cos 2\pi\alpha} \\[2mm] \frac{\zeta}{\lambda} &= \frac{2}{\pi} - \frac{2}{\pi} \frac{1 + A \cos 2\pi\alpha}{1 + A^2 + 2A \cos 2\pi\alpha} \end{aligned} \right\} \qquad (2\text{-}64)$$

where A depends on the wave steepness δ as

$$A = \frac{2}{\pi\delta} \left\{ \left(1 + \frac{\pi^2\delta^2}{4} \right)^{1/2} - 1 \right\}. \qquad (2\text{-}64')$$

The phase speed is given by

$$c = \left(\frac{2\pi}{\lambda} \frac{S}{\rho_w} \right)^{1/2} \left(1 + \frac{\pi^2\delta^2}{4} \right)^{-1/4}. \qquad (2\text{-}65)$$

As shown in Fig. 2-3, the wave profile is almost sinusoidal when the amplitude is small, but as the steepness δ increases, the crest is rounded and the trough pointed. The maximum possible steepness is 0.730. A very high frequency component of wind waves is observed to have a shape similar to these waves.

2.3.3 Mass transport, momentum and energy of surface waves

The Lagrangian velocity u_1 of a water particle is, to the first order approximation, equal to the Eulerian velocity. For deep water waves, the velocity components in the vicinity of (x_0, z_0) are expressed by

$$\left. \begin{array}{l} u_1 = \sigma a e^{kz_0} \sin(kx_0 - \sigma t) \\[2mm] w_1 = \sigma a e^{kz_0} \cos(kx_0 - \sigma t) \end{array} \right\} . \qquad (2\text{-}66)$$

Amplitudes of u_1 and w_1 are a function of z_0, which is the average position in the z coordinate of the particle. The mean of u_1 or w_1 is zero. However, the Lagrangian velocity to second order is not zero, that is

$$\left. \begin{array}{l} \bar{u}_1 = k^2 a^2 c e^{2kz_0} \\[2mm] \bar{w}_1 = 0 \end{array} \right\} . \qquad (2\text{-}67)$$

Therefore, the particle motion cannot be closed for a finite amplitude wave, but there is a slight forward drift on the average in the direction of wave propagation with a slight net transport of mass. This mean value of the Lagrangian velocity is

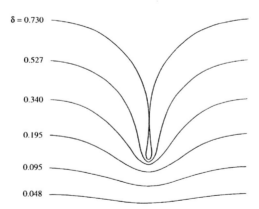

$\delta = 0.730$

0.527

0.340

0.195

0.095

0.048

Fig. 2-3. Streamline or wave form of progressive capillary waves (after Crapper, 1957). The parameter $\delta = H/\lambda$.

called *mass transport velocity*, and the generated flow is called the *mass transport* or the *wave current*, or the *Stokes drift*. Although for irrotational waves an average of the Eulerian velocity below the trough level is zero everywhere, the velocity given by Eq. (2-66) is a function of depth. From a Lagrangian point of view, in the case of finite amplitude, a particle above its mean depth moves faster in the forward direction than it does in the backward direction beneath its mean depth, thus producing a net mass transport.

The mass flux density M due to u_1 is given by

$$M = |M| = \rho_w \int_{-\infty}^{0} \bar{u}_1 \, dz_0 = \frac{1}{2} \rho_w \sigma a^2 . \tag{2-68}$$

Since the mass flux density is equal to the momentum, M is nothing but the mean *wave momentum* per unit horizontal area. As described above, the wave momentum is carried by particles moving above the level of the wave trough at each instant.

The mean *wave energy* per unit surface area of deep water waves is defined by

$$E = \rho_w \overline{\int_{-\infty}^{\zeta} u^2 \, dz} . \tag{2-69}$$

The approximation of E to the second order (for an infinitesimal wave) is given by

$$E = \rho_w \int_{-\infty}^{0} \overline{u^2} \, dz = \frac{\rho_w \sigma^2 a^2}{2k} . \tag{2-70}$$

For gravity waves, $Sk^2/\rho_w \ll g$, and then

$$E = \frac{1}{2} \rho_w g a^2 = \rho_w g \overline{\zeta^2} \tag{2-71}$$

while for capillary waves, $Sk^2/\rho_w \gg g$, and then

$$E = \frac{1}{2} Sk^2 a^2 = S\overline{(\nabla \zeta)^2} . \tag{2-72}$$

The mean kinetic and potential energies are half of these values, respectively. The wave energy propagates at its *group velocity*

$$c_g = \frac{d\sigma}{dk} . \tag{2-73}$$

For deep water waves of infinitesimal amplitudes,

$$c_g = \left(1 - \frac{1}{2}\frac{\lambda^2 - \lambda_m^{\ 2}}{\lambda^2 + \lambda_m^{\ 2}}\right)c \qquad (2\text{-}74)$$

where λ_m is the wavelength of the waves at the minimum phase speed. For a particular case of gravity waves, $c_g = c/2$, and for capillary waves, $c_g = 3c/2$.

Combining Eqs. (2-59), (2-68) and (2-71) leads to the relationship between the wave momentum and wave energy as follows:

$$M = \frac{E}{c}\frac{k}{k}. \qquad (2\text{-}75)$$

For finite amplitude waves, the kinetic energy E_k is greater than the potential energy E_p, due to the existence of mass transport. Only the kinetic energy is related to the momentum,

$$M = \frac{2E_k}{c}. \qquad (2\text{-}76)$$

The momentum flux M_f at the wave speed c is changed from the linear theory case ($=E/2$) to

$$M_f = 4E_k - 3E_p. \qquad (2\text{-}77)$$

And the energy flux E_f is changed from the linear theory case ($=E_c/2$) to

$$E_f = \left(3E_k - 2E_p\right)c. \qquad (2\text{-}78)$$

The speed of energy propagation V_E, which corresponds to c_g for linear waves, is now (Longuet-Higgins, 1975)

$$V_E = \frac{3E_k - 2E_p}{E_k + E_p}c \qquad \left(> \frac{1}{2}c\right). \qquad (2\text{-}79)$$

The mass transport of finite amplitude gravity waves, the Stokes drift, can be regarded as a Lagrangian drift. However, if we consider the vorticity for its mean velocity, we would obtain, from Eq. (2-67),

$$\eta = -2\sigma a^2 k^2 e^{2kz_0}. \qquad (2\text{-}80)$$

This coincides with the vorticity of the *Gerstner wave*, which is an exact analytical solution for finite amplitude waves as a rotational flow, in its absolute value to the second order, with the opposite sign. The Stokes wave is locally irrotational at each instant, but from another point of view it has a vorticity expressed by Eq. (2-80) on an average. On the other hand, the Gerstner wave has no mass transport, but it has the same amount of local vorticity in the opposite sign at each instant. In this sense, the Gerstner wave can be viewed as the Stokes wave with the negative mass transport added. That is, although it is difficult mathematically to transform the Stokes wave to the Gerstner wave or vice versa, the two waves may be interpreted as a sort of twins in a physical sense as described above.

2.3.4 Wave–current interactions

As discussed in the previous section, a finite amplitude wave has a mass transport and momentum. Therefore, if a wave train and a current, which varies with time or space, are coexistent, a nonlinear interaction between the waves and the current occurs: mass, momentum and kinetic energy are transferred between these modes. The result appears as the change in the steepness of the waves when they propagate to a current area, the change in the mean sea level when swell moves towards a shoreline, or steepening of short waves at the crest of long waves.

In a system where both waves and a current exist, conservation of the total mass transport $\tilde{M}$ per unit breadth is related to the averaged total mass transport velocity $\tilde{U}$ as

$$\frac{\partial}{\partial t}\tilde{M}_\alpha + \frac{\partial}{\partial x_\beta}\left(\tilde{U}_\alpha\tilde{M}_\beta + S_{\alpha\beta}\right) = T_\alpha. \qquad (2\text{-}81)$$

In this equation, the excess momentum flux density tensor $S_{\alpha\beta}$ has appeared because of the existence of the non-stationary motion. α and β are two components (1,2) in the horizontal direction, and $S_{\alpha\beta}$ can be expressed as

$$S_{\alpha\beta} = \overline{\int_{-h}^{\zeta}\left(\rho_w u_\alpha' u_\beta' + p\delta_{\alpha\beta}\right)dz} - \frac{1}{2}\rho_w g\left(h+\overline{\zeta}\right)^2\delta_{\alpha\beta} - \frac{M_\alpha M_\beta}{\rho_w\left(h+\overline{\zeta}\right)} \qquad (2\text{-}82)$$

where u' is the speed due to the waves, M the mass transport of waves, h the water depth, and $\overline{\zeta}$ is the mean displacement of the sea level caused by the interaction between the waves and the current. T_α in Eq. (2-81) is the pressure gradient force per unit horizontal area, which is caused by the slope of the free surface,

$$T_\alpha = -\rho_w g\left(h+\overline{\zeta}\right)\frac{\partial\overline{\zeta}}{\partial x_\alpha}. \qquad (2\text{-}83)$$

The second term on the right-hand side of Eq. (2-82) is one that would cancel out with an integral of the pressure in the first term.

The balance of the wave energy E, if no wave energy is dissipated, is described by

$$\frac{\partial E}{\partial t} + \frac{\partial}{\partial x_\alpha}\left[E\left\{U_\alpha + \left(c_g\right)_\alpha\right\}\right] + S_{\alpha\beta}\frac{\partial U_\beta}{\partial x_\alpha} = 0, \qquad (2-84)$$

with U the velocity of the mean current. Longuet-Higgins and Stewart (1964) named $S_{\alpha\beta}$ the *radiation stress*: it represents the excess momentum flux due to the presence of waves. If we consider the fluctuation as turbulence, not as waves, then $S_{\alpha\beta}$ corresponds to an integrated Reynolds stress. Therefore, the last term in Eq. (2-84) can be regarded as the work done to the mean flow by the existence of the fluctuating motion.

As a simple example, let us take an axis in the direction of propagation of a single wave train. In this case we have

$$S_{\alpha\beta} = \begin{pmatrix} E\left(\dfrac{2c_g}{c} - \dfrac{1}{2}\right) & 0 \\ 0 & E\left(\dfrac{c_g}{c} - \dfrac{1}{2}\right) \end{pmatrix}. \qquad (2-85)$$

For deep water waves, the only non-zero component is in the direction of wave propagation: $S = E/2$. For pure capillary waves, $S = 3E/2$.

2.3.5 Wave–wave interactions

If a wave is long enough, it can be approximated by a current as seen in the previous section. But if two waves are both short, we are to consider the *wave–wave interactions*. Waves of infinitesimal amplitude, however complex they are, can be treated as a superposition of component waves traveling independently of one another, because their particle motions are small compared with the propagation speed, making a linear approximation possible. However, the waves in the ocean are of finite amplitude. Even if a wave train is decomposed into wave component modes, the nonlinear terms, which were neglected in the linear theory, now generate interactions among the wave components. The interaction, though weak, causes energy exchange among the modes. That is, Fourier coefficients of the modes vary with time.

Weak nonlinear interactions can be treated as perturbations of superposed linear solutions. In the interaction to the second order of wave steepness, energy exchange by resonance can occur among the three wave components, which satisfies the relationship

$$k_1 \pm k_2 = k_3 \\ \sigma_1 \pm \sigma_2 = \sigma_3 \biggr\} . \qquad (2\text{-}86)$$

It should be noted that the solution can exist only when the dispersion relationship between σ_i and k_i holds. Because of this restriction, there is no solution for surface gravity waves; there are solutions for surface gravity-capillary waves, internal waves in a stratified fluid, or two surface gravity waves and one internal wave.

For surface gravity waves, it is known that a resonant interaction to the third order among four waves is possible if the following conditions are satisfied:

$$k_1 + k_2 = k_3 + k_4 \\ \sigma_1 + \sigma_2 = \sigma_3 + \sigma_4 \biggr\} \qquad (2\text{-}87)$$

where $\sigma_i^2 = gk_i$ and $i = 1, 2, 3,$ or 4.

Although energy is exchanged among the waves, the total energy is conserved. Since surface gravity waves start to interact at their third order, the wave–wave interaction is rather weak and the energy exchange among the wave components is gradual.

2.3.6 Spectral representation of waves

In a model where ocean waves are treated as a collection of component waves, the wave field is expressed as the superposition, in random phases, of mutually independent infinitesimal waves, propagating in various directions and having a certain distribution of wave heights, with a continuous frequency distribution. They are to be expressed in an energy spectral form.

Assuming statistically uniform and stationary conditions, we express a sea level record ζ at a point as

$$\zeta(t) = \sum_n a_n \cos(\sigma_n t + \varepsilon_n). \qquad (2\text{-}88)$$

When ε_n is random, the mean square of the sea level is

$$\overline{\zeta^2} = \frac{1}{t_1} \int_0^{t_1} \zeta^2 dt = \sum_n \frac{1}{2} a_n^2 . \qquad (2\text{-}89)$$

This can be written as

$$\overline{\zeta^2} = \int_0^\infty \phi(\sigma) d\sigma = E , \qquad (2\text{-}90)$$

where the *spectral density* $\phi(\sigma)$ gives the one-dimensional *frequency spectrum*. This formula will give energy density if multiplied by $\rho_w g$.

If the direction of wave propagation is taken into account and when the sea level record at the position r is written as

$$\zeta(r,t) = \sum_n a_n \cos(k_n \cdot r - \sigma_n t + \varepsilon_n), \qquad (2\text{-}91)$$

then

$$\overline{\zeta^2} = \int_0^{2\pi}\int_0^{\infty} \psi(k,\theta)k\,dk\,d\theta = \int_0^{2\pi}\int_0^{\infty} \Phi(\sigma,\theta)\sigma\,d\sigma\,d\theta, \qquad (2\text{-}92)$$

where $\psi(k, \theta)$ and $\Phi(\sigma, \theta)$ are *directional (two-dimensional) energy spectra* with respect to the wave number and frequency, respectively, and θ is the direction. In the gravity wave region where the dispersion relationship $\sigma^2 = gk$ holds, the wave number spectrum is related to the wave frequency spectrum as

$$\Phi(\sigma,\theta) = \frac{2}{g}\bigl[k\psi(k,\theta)\bigr]_{k=\frac{\sigma^2}{g}}. \qquad (2\text{-}93)$$

This is related to the one-dimensional spectrum as

$$\phi(\sigma) = \int_0^{2\pi} \Phi(\sigma,\theta)\sigma\,d\theta. \qquad (2\text{-}94)$$

We can write, for convenience sake,

$$\Phi(\sigma,\theta) = \phi(\sigma)G(\sigma,\theta), \qquad (2\text{-}95)$$

where $G(\sigma, \theta)$ is a distribution function showing how the energy is distributed according to the propagation direction. It is called the *directional distribution function*.

To calculate the spectrum from a time series of sea level records, the method most widely used is FFT (fast Fourier transform): Fourier coefficients of the records are first obtained directly by the Fourier transform, and then squared, which provides the estimated energy after an averaging process is made.

The characteristics of the real ocean wave spectra will be discussed in Section 2.4.

2.3.7 Initial generation of wind waves on a still water surface

If wind waves are regarded, as done in early years, as a superposition of irrotational waves, no forces other than varying pressure distributed over the sea surface could generate waves. This concept was formulated in the frequently quoted works of two scientists. Phillips (1957) presented a resonant mechanism for the wave generation on a still water surface, whereas Miles (1957) proposed an instability theory on the further development of generated waves.

Phillips' theory concerns resonance between air turbulence pressure fluctuation and water waves underneath. As the wind with turbulence starts to blow on a still water surface, its varying pressure field also travels, and a *resonant mechanism* is triggered to generate waves in such a condition that the speed and wavelength of the pressure variation coincide with the phase speed and wavelength of a water wave. If the average wind speed U at a height of the order of the wavelength is greater than the wave speed c, the wave is assumed to propagate in two directions of $\cos^{-1}(c/U)$. The wave energy grows in proportion to time in this mechanism.

Miles' theory deals with the instability in the coupling of an infinitesimal sinusoidal wave field on a water surface with a shear flow of air over it. The instability requires the pre-existence of some infinitesimal waves to make the wave amplitude of a certain frequency increase further. A characteristic feature is that a *critical height* exists where the speed of the mean flow U and the wave speed are equal. That is, if the coordinate system is taken as moving with the wave, a layer of no averaged motion with respect to the coordinate is present at that height (see Fig. 2-4). Pressure variation of the air over the waves is also sinusoidal. Assuming that Bernoulli's theorem holds approximately, if the pressure variation and the stream function are either 180 degrees out of phase (or just in phase) as shown in Figs. 2-5(a) and (b), the perturbed velocity fluctuations (u, v) induced by the wave will provide no Reynolds stress ($\overline{uv} = 0$). However, other cases with some phase shift, as expressed in Fig. 2-5(c), Reynolds stress $-\overline{uv}$ induced by the waves will exist. According to Miles' mechanism, this phase shift is caused originally by the diffusion of vorticity, due to viscosity or turbulence, in the presence of a curvature at the critical height. If we take a close look at the neighborhood of the critical height, a closed streamline can be considered as seen in Fig. 2-5(α). Along the closed stream line, when an air

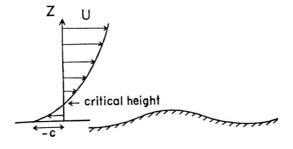

Fig. 2-4. Existence of the critical height.

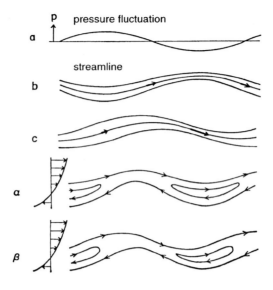

Fig. 2-5. Concept of Miles' mechanism. (See the text.)

particle passes above the critical height, negative vorticity is lost, while negative vorticity is gained when it passes below, due to the shape of the mean air flow profile. Then the closed streamline becomes asymmetric and thus unclosed streamlines are deviated, too, as shown in Fig. 2-5(β). Accordingly, the efficiency of wave development depends on the difference in vorticities between both sides of the critical height, or the curvature of the mean velocity of air near this layer: $U_c'' = d^2 U_c/dz^2$. The wave growth is exponential, as is the case with the *instability mechanism*.

Kawai (1979) showed, by experiments and theory, that the generation mechanism of the initial wavelets on a still water surface by wind is instability of the coupled shear flow of the viscous fluids of air and water, producing a selective amplification of disturbances of the frequency at the maximum growth rate. The growth rate corresponds to a modified Miles' mechanism after viscosity is taken into account. This is accepted today as the generation mechanism of wind waves, although more researches are being conducted to test the theory further.

However, as soon as initial wavelets are generated, they are immediately accompanied by turbulence, and the actual wave development progresses with turbulence. At a certain stage of transition from initial wavelets to wind waves, particular *rhombic waves* are generated (Ebuchi *et al.*, 1987). This result does not agree with McLean's two-dimensional instability theory. It is probably related to unstable phenomenon of water waves in a shear flow, but it is not solved theoretically yet. In the developing process a spectral peak is known to shift to lower frequencies discretely (Hatori, 1984). These topics, which should be categorized somewhere between this section and the following section, are still under active studies all over the world as questions of water waves under special influence by the wind.

2.4 WIND WAVES AND COUPLED BOUNDARY LAYERS OF ATMOSPHERE AND OCEAN

The atmosphere and the oceans interact with each other locally through physical processes related to wind waves. In this sense, it is necessary to understand wind-wave phenomena as a basis for parametrizing ocean–atmosphere exchange processes of momentum, CO_2 and other quantities.

2.4.1 Characteristics of wind waves

"Swell", which is a state of gravity waves when they have escaped the influence of the generating wind, are found universally in the ocean and can be described quite properly as a superposition of component waves. However, pure wind waves in their generation areas have special characteristics that may not be expressed merely as a combination of gravity and capillary waves. In short, *wind waves* are a particular fluid phenomenon that is generated at the boundary of two viscous fluids of air and water moving at different speeds, and that includes local wind drift, wave breaking and turbulence intrinsically. From now on, the author will use a term *windsea* to express this particular phenomenon. Readers can also refer to Toba (1998) and Csanady (2001).

As described in the previous section, windsea is, from its initial generation stage, a phenomenon particular to viscous fluids. In addition to the interactions among component waves as discussed earlier, many elements that are combined with one another in a strongly nonlinear manner produces windsea. Among them are, the air flow separation related to the presence of *individual wave* crests of windsea, reattachment of the separated flow onto the wave surface, the resultant shear and pressure stresses of the wind that are distributed along each phase of the individual waves, the local wind-drift shear flow of the water side, turbulence due to the shear flow, and wave breaking.

Breaking of windsea in the water side corresponds to the air-flow separation of the air side. Besides the obviously visible wave breaking with the air-bubble entrainment, there is wave breaking in which no air is entrained and so usually not visible. Banner and Phillips (1974) termed this kind of breaking *incipient breaking*. Corresponding to *ordered motion* in the boundary layer of the air side, the bubbles are transported downward together with water particles.

Through these strongly nonlinear processes, a special self-adjustment mechanism appears to work in the wind-wave field. Windsea has a similarity structure in the energy spectrum, and the 3/2-power law as a macroscopic similarity law which is directly associated with it. The continuities of surface velocity, turbulence structure and momentum transfer in the turbulent boundary layers above and below the wind-wave surface must be achieved by a series of these self-adjustment processes.

2.4.2 Similarity laws of windsea—Windsea in local equilibrium with wind

The energy spectrum of pure windsea has a similarity structure with a conspicuous peak at a certain frequency flanked with steep energy decreasing ranges. This allows us to treat the windsea field by representative parameters such as the spectral peak frequency. Since Sverdrup and Munk (1947), concepts of *significant wave height* and *significant wave period*[1] have been widely used. The significant wave period approximately corresponds to the period of waves at the spectral peak frequency. How windsea develops as the fetch and duration of the wind varies has been formulated empirically, e.g., by Wilson (1965) as follows:

$$H = 0.30\left[1 - \left(1 + 0.004X^{1/2}\right)^{-2}\right], \tag{2-96}$$

$$T = 1.37\left[1 - \left(1 + 0.008X^{1/3}\right)^{-5}\right], \tag{2-97}$$

where H, T and X are nondimensional variables defined by

$$H = \frac{gH_s}{U_{10}^2}, \quad T = \frac{gT_s}{2\pi U_{10}}, \quad X = \frac{gx}{U_{10}^2} \tag{2-98}$$

and H_s is the significant wave height, T_s the significant wave period, U_{10} the wind speed at the 10-m level, and x the fetch. If the fetch is short ($X < 10^3$),

$$H = 0.0024X^{1/2}, \tag{2-99}$$

$$T = 0.0548X^{1/3}. \tag{2-100}$$

If the fetch is long enough, the windsea asymptotically approaches a fully developed, saturated state.

In Eqs. (2-99) and (2-100) or formulae by Mitsuyasu (1971) similar to these, H and T are proportional to $X^{1/2}$ and $X^{1/3}$ respectively. If X is eliminated from these equations, a simple relationship $H^2 \propto T^3$ is expected. As a matter of fact, even if we use rather complex empirical formulae such as Eqs. (2-96) and (2-97), which cover broader regions including the state approaching saturation, the relationships between H and T do have that form in a wide range of X as shown

---

[1]For a wave height record with a specified duration of time (for instance, 1024 seconds), the number of individual waves is counted, and then the mean height of the highest one-third of the waves is called the significant wave height; the mean period of the same waves is called the significant wave period.

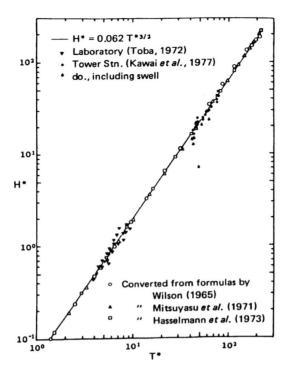

Fig. 2-6. The 3/2-power law of wind waves: Eq. (2-101) with a data set (after Kawai *et al.*, 1977).

in Fig. 2-6. It is seen that results from other experiments and observations also support these relationships. It was expressed, by using the air friction velocity u_*, as the *3/2-power law* (Toba, 1972)

$$H^* = BT^{*3/2} \qquad (2\text{-}101)$$

where

$$H^* = \frac{gH_s}{u_*^2}, \quad T^* = \frac{gT_s}{u_*}, \quad B = 0.062 \qquad (2\text{-}102)$$

and B is an empirical constant. It is to be noted that B is a statistical average value, and it shows perturbation of about ±20% due to the wind fluctuation.

It is often difficult to define the fetch x in the real ocean, since the wind field always varies. Nevertheless Eq. (2-101) holds when swells are not significant. It should be stressed that this similarity law of windsea is of very simple form containing u_*. The interpretation is that H_s and T_s do not vary independently, but are combined with each other through the local wind field which is represented by u_*, and that the wind and the windsea are in *local equilibrium*.

The 3/2-power law can be rewritten, using the *significant wave steepness δ* (defined as the ratio of the significant wave height to the wavelength of the corresponding waves), as a function of the *wave age c_p/u_**, where c_p is the phase speed of the significant waves (Bailey *et al.*, 1991), see Fig. 2-7,

$$\delta = \frac{B(2\pi)^{1/2}}{\left(c_p / u_*\right)^{1/2}} \left(1 + \frac{u_s / u_*}{c_p / u_*}\right)^{-2/3} , \quad \frac{u_s}{u_*} = 0.21 . \qquad (2\text{-}103)$$

If we have only the 3/2-power law, δ would become infinitely large as the wave age approaches zero. In the real sea, however, the local wind drift u_s exists and the correction term containing u_s is included in Eq. (2-103), which has a finite limit of δ for very young sea (Tokuda and Toba, 1982).

Since the 3/2-power law is a form for significant waves, it agrees with data for pure windsea with little swell energy. However, when swell components are large, data points on Fig. 2-6 would shift to the lower right portion to the straight line, and the 3/2-power law indicates the upper limit of the data-point distribution. If the wind suddenly weakens as is the case after a typhoon or a low pressure has passed, u_* suddenly becomes small while the wave quantities H_s and T_s hardly change immediately, so that the data points would shift to the upper right direction along the slope of $H \propto T^2$. Generally speaking, when the wind is weak, or H_s is small, swell predominates and the data points deviate from the 3/2-power law. However, when H_s exceeds 4 m, the local equilibrium associated with the 3/2-power law stands fairly well (Ebuchi *et al.*, 1992). Even when the significant

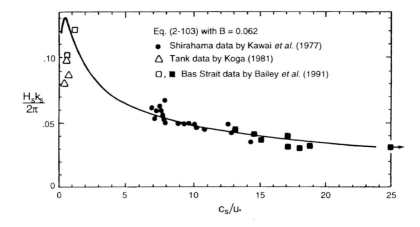

Fig. 2-7. Significant steepness of wind waves as a function of wave age with some observed data (after Bailey *et al.*, 1991).

waves do not agree with the local equilibrium formula in the presence of swells, the high frequency range of windsea spectra is regarded as satisfying the local equilibrium formula.

As a supplementary note, the statistics of significant waves and significant wave heights are worth mentioning. The probability density function $p(a)$ for the amplitude a of a random fluctuation in a narrow spectral band is given by the Rayleigh distribution

$$p(a) = \frac{2a}{\overline{a^2}} \exp\left(-\frac{a^2}{\overline{a^2}}\right).$$ (2-104)

The relationships among the mean square amplitude $\overline{a^2}$, and the mean amplitude $\overline{a}$ and the significant wave height H_s are given by (Longuet-Higgins, 1952)

$$\frac{\overline{a}}{\sqrt{\overline{a^2}}} = \frac{\sqrt{\pi}}{2} = 0.886, \quad H_s = 1.60\overline{H}, \quad \overline{H} = 2\overline{a}, \quad H_s = 2.83\sqrt{\overline{a^2}}.$$ (2-105)

2.4.3 Windsea spectra

The spectral form of the high frequency side of a one-dimensional self-similar spectrum, which would coincide, after integration, with the 3/2-power law Eq. (2-101), is expressed by

$$\phi(\sigma) = \alpha_s g_* u_* \sigma^{-4}, \quad \sigma > \sigma_p,$$ (2-106)

where $\phi(\sigma)$ is the energy density, σ the angular frequency, σ_p the spectral peak angular frequency, g_* the extended acceleration of gravity to include the surface tension S,

$$g_* = g\left(1 + \frac{Sk^2}{\rho_w g}\right),$$ (2-107)

k the wavenumber, and ρ_w the water density (Toba, 1973). For pure gravity waves, g_* can be replaced by g. Spectra of actually observed windsea shows this proportionality to $gu_*\sigma^{-4}$ on the main part of the high frequency side. The spectral range that can be expressed by Eq. (2-106) will be called the *equilibrium range*.

Assuming that gravity causes wave breaking and dimensionality, Phillips (1957) first proposed

$$\phi(\sigma) = \alpha g^2 \sigma^{-5}$$ (2-108)

for the *equilibrium range*. This was widely accepted in early years, but many observations agreeing with Eq. (2-106) rather than Eq. (2-108) were reported, and Phillips (1985) re-proposed the new Eq. (2-108) type of spectrum.

The value of α_s in Eq. (2-106), which is approximately $(6-12) \times 10^{-2}$ including an error for the measurement of u_*, apparently changes depending on the response process of the windsea to wind fluctuations. Figure 2-8 shows some examples of observed spectra made at an oceanographic tower station. From the condition that upon integration the level of Eq. (2-106) should agree with Eq. (2-101), α_s can be estimated as 0.096 corresponding to $B = 0.062$ (Joseph *et al.*, 1981).

Mitsuyasu and Honda (1974) reported that, in the very high frequency region of 10–50 Hz where both gravity and capillary effects are important, the spectral density is proportional to the 2.5 power of u_*, not to u_* itself. This topic will be discussed further in Chapter 5 in relation to satellite observations.

In recent years researches on the two-dimensional spectrum using optical methods (such as a laser inclinometer for the water surface), have progressed. Figure 2-9 shows a result of Jähne and Riemer (1990). It is seen that the high frequency part of the spectrum is dependent on U_{10}. Note that the spectral form $B(f)$ normalized by $u_* f^{-4}$ is used for the ordinate, so that all values converge at the region near 3 Hz which corresponds to Eq. (2-106). Since the irregularities of these high frequencies appear and disappear almost instantly (Ebuchi *et al.*, 1987), the phenomenon is seemingly more directly associated with turbulence rather than with waves.

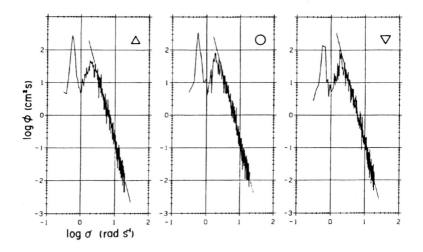

Fig. 2-8.   Observed one-dimensional frequency spectra of wind waves in the sea. Ensemble averages of ten of the raw spectra for increasing winds (triangle), constant winds (circles), and decreasing winds (inverse triangle). A peak at the left side indicates swell (after Toba *et al.*, 1988).

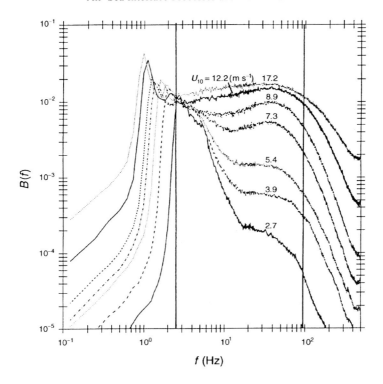

Fig. 2-9. Frequency spectra of wind waves for high frequency ranges measured with a
laser slope gauge. $B(f)$ is the value normalized by $u_* f^{-4}$ (after Jähne and Riemer, 1990).

A large number of studies on empirical forms of one-dimensional spectra
and characteristics of their development have been accumulated (cf. Huang *et al.*,
1990). Pierson and Moskowitz (1964) proposed

$$\phi(f) = \alpha g^2 (2\pi)^{-4} f^{-5} \exp\left\{-\frac{5}{4}\left(\frac{f}{f_p}\right)^{-4}\right\}, \quad \alpha = 8.1 \times 10^{-3}. \quad (2\text{-}109)$$

This form includes roundness near the spectral peak and a decreasing slope on the
low frequency side. Hasselmann *et al.* (1973) proposed, based on data from the
JONSWAP project, a modification of Eq. (2-109) by multiplying it by a *peak
enhancement* function

$$\gamma^A, \quad A = \exp\left\{-\frac{\left(f - f_p\right)^2}{2C^2 f_p^2}\right\}, \quad (2\text{-}110)$$

where $\gamma = 3.3$, $C = 0.07$ ($f \leq f_{\mathrm{p}}$), $C = 0.09$ ($f > f_{\mathrm{p}}$), and α was assumed to decrease as the nondimensional fetch X increases:

$$\alpha = 0.076X^{-0.22} . \qquad (2\text{-}111)$$

This modification accentuates the energy concentration around the peak.

For the directional distribution function in Eq. (2-95), Longuet-Higgins *et al.* (1961) proposed the formula

$$G(f,\theta) = G'(S)\cos^{2s}\frac{(\theta - \bar{\theta})}{2} , \qquad (2\text{-}112)$$

where $\bar{\theta}$ is the mean direction of the waves, $G'(S)$ is the normalization function so that $\int_0^{2\pi} G(f,\theta)d\theta = 1$, and S is a parameter indicating the directional concentration. Based on their observations, Mitsuyasu *et al.* (1975) proposed a standard form:

$$\left. \begin{array}{ll} \dfrac{S}{S_{\mathrm{p}}} = \left(\dfrac{f^*}{f_{\mathrm{p}}^*}\right)^{-2.5} , & f^* \geq f_{\mathrm{p}}^* \\[4mm] \dfrac{S}{S_{\mathrm{p}}} = \left(\dfrac{f^*}{f_{\mathrm{p}}^*}\right)^{5} , & f^* \leq f_{\mathrm{p}}^* \end{array} \right\} \qquad (2\text{-}113)$$

where $f^* = u_* f / g$.

Windsea which is generated in a wind-wave tank of short fetches is more regular, the spectral slopes on both sides of the peak have a gradient of ±9th or ±10th power of σ, and the second and the third peaks appear at higher harmonics. Since these higher harmonic components are due to the distortion of individual waves, they propagate at approximately the same speed as the waves of the spectral peak (Masuda *et al.*, 1979; Mitsuyasu *et al.*, 1979).

Non-stationary and non-uniform characteristics of windsea are being studied by using new techniques such as Hilbert transform and wavelet analyses (cf. Huang, 1998).

2.4.4 *Coupling of windsea and the ocean–atmosphere boundary layers*

A theoretical study by Zakharov and Filonenko (1966) showed that the water wave spectrum has a region proportional to σ^{-4} due to wave interactions. However, the experimental results as described above indicate that there exists the equilibrium range where the spectral form is proportional not only to σ^{-4} but also to u_*, and we can say that its integral form leads to the relationship between the nondimensional significant wave height H^* and period T^* as shown in Eq. (2-101). The existence of the spectral range proportional to g and u_* had been predicted dimentionally by Kitaigorodskii (1961). However, no purely theoretical work had succeeded in deriving why the energy level of the equilibrium range is proportional to u_*, without relying on the experimental results. Tulin (1996) proposed to explain the 3/2-power law by simultaneous consideration of the change rates of energy and momentum of windsea, resulting from wave breaking, keeping the wave momentum and energy relation (2-75). Thie leads to the necessary downshifting of windsea frequency to satisfy the 3/2-power law, keeping a balance between the wind input and dissipation by wave breaking.

Equation (2-101) corresponds also to the fact that individual waves have the same wave current in common,

$$u_0 = 2\pi^3 B^2 u_* , \qquad (2\text{-}114)$$

and that this wave current is proportional to u_* (Toba, 1988). It can be theoretically shown (Bye, 1988) that when the high frequency spectrum of windsea is of the type $u_*\sigma^{-4}$, then the wave current is proportional to u_*.

The air flow over laboratory windsea has a range of logarithmic boundary layer where the friction velocity u_* can be considered constant vertically. This layer corresponds to the turbulent boundary layer over rough solid surfaces discussed in Section 1.1.2. In this logarithmic profile range, ordered motions exist, as in the case over solid surfaces (Kawamura and Toba, 1988). However, the windsea case differs from the solid surface one, in that the water has also a turbulent boundary layer below the surface whose structure is very much like that of the air boundary layer except that the direction is opposite. Experimental results show the following relationship among the friction velocities of air u_* and water u_{*w}, the turbulent velocities of air u_a', and of water u_w' in which the water particle motion due to waves was excluded:

$$u_* \propto \left(\overline{u_a'^2}\right)^{1/2} \propto \left(\overline{u_w'^2}\right)^{1/2} \propto u_{*w} . \qquad (2\text{-}115)$$

This indicates that the windsea creates a combined boundary layer of air and water where the characteristic velocity is u_*. Therefore, together with Eq. (2-114), the 3/2-power law of windsea expresses the fact that the water-wave element of the windsea is combined with the turbulent intensities of the air and the water boundary layers (Yoshikawa et al., 1988).

Just below the windsea, there exists a special boundary layer (*downward bursting turbulent boundary layer*; DBBL) whose K_M is greater than that of the logarithmic boundary layer on a solid surface as expressed in Eq. (2-31). As subsequent experimental results, Toba and Kawamura (1996) showed that the depth of this special boundary layer is approximately five times as large as H_s, from the tiny invisible breaking immediately after the initial generation of windsea to the large breaking of ocean waves. See also a review by Toba (1998).

2.5 PARAMETRIZATION OF AIR–SEA BOUNDARY PROCESSES

The parametrization of the air–sea boundary processes should be an important element for modeling ocean–atmosphere interactions. This section deals with questions for such parametrizations.

2.5.1 What is parametrization?

As discussed previously, there are two boundary layers above and below the sea surface, which are coupled by windsea. Between the atmosphere and the ocean, the exchange of momentum, heat, water vapor, carbon dioxide, salt particles, etc. proceeds by turbulent motions directly connected with the windsea. In addition to the sensible and latent heat exchanges, there are incoming and outgoing heat by short and long wave radiation at the sea surface.

For instance, the content of the momentum flux in the air boundary layer is the turbulent transfer expressed by Eq. (2-26) or the left-hand side of Eq. (1-15) in Chapter 1. However, measuring the product of the fluctuating velocities directly for estimating flux will require high technical skills and much laborious work: for example, we may have to set up, with great care, an ultrasonic anemometer on a measuring platform such as a fixed observation tower. This is not feasible for routine observations everywhere. Nonetheless, what is necessary for a numerical model to forecast large scale phenomena or for studies on large-scale ocean–atmospheric interactions is more macroscopic quantities, not detailed structures of the fluxes.

There are various degrees of scales in phenomena of the ocean and the atmosphere: from wind waves to general circulation in the ocean, for instance. To model each of these, we need the most appropriate approximation or modification of the equations. However, a larger-scale phenomenon is affected by smaller-scale processes. Therefore, if smaller-scale processes can be brought into a larger-scale model as appropriate parameters, that is, if parametrization of smaller-scale processes is made possible by more macroscopic quantities, then we can connect models dealing with phenomena of different scales. We can then study interactions among processes with various scales. This is the concept of parametrization.

2.5.2 Bulk formulas for sea surface fluxes

Examples of substantial processes at the sea surface include fluctuating fields, by turbulence, of wind, temperature, water vapor density, and surface wave field with windsea. Measurable quantities are, first, (1) fields of varying physical quantities. Next, as simpler quantities, we have (2) vertical distributions (profiles) of average quantities, and as further simple quantities, (3) mean values of various quantities at a standard height (say, 15-minute average values measured at the 10-m height), sea surface temperature (not skin layer but bulk sea water temperature), salinity, significant wave height, and so forth. Those in item (3) are representative quantities. If the various fluxes can be expressed in terms of these values, they can be conveniently used as boundary values in modeling. Such expressions are called bulk formulas.

An example is the *wind stress* τ at the sea surface expressed, considering the dimensions, by

$$\tau = -\rho \overline{u'w'} = \rho C_D U_{10}^2 , \qquad (2\text{-}116)$$

where U_{10} is the mean wind speed at the 10-m height, ρ the air density, and C_D is the *drag coefficient* of the wind on the sea surface. The nondimensional coefficient C_D is not necessarily a constant, but is to be known only empirically. It should be given as a function of various parameters by detailed observational studies. The bulk formulas for the heat and water vapor fluxes are, in a similar manner, written as

$$H = \rho C_p C_H U_{10}\left(T_s - T_{10}\right), \qquad (2\text{-}117)$$

$$E = \rho C_E U_{10}\left(q_s - q_{10}\right), \qquad (2\text{-}118)$$

where C_H and C_E are the *bulk transfer coefficients* for sensible heat and water vapor, respectively, H the upward sensible heat transfer, E the evaporation from the sea surface, C_p the specific heat at constant pressure, T_s here the sea surface temperature, and q_s the equilibrium specific humidity at the sea surface. q_s is a function of water temperature and salinity and the effect of salinity on vapor pressure is about 1.7%.

For a neutral stratification, combination of the definition of C_D, Eq. (2-116), the definition of friction velocity u_*, Eq. (2-20), and the logarithmic law of the wind profile, gives

$$u_* = \sqrt{C_D}\,U_{10} \qquad (2\text{-}119)$$

and

$$C_D = \left(\frac{1}{k}\ln\frac{z_{10}}{z_0}\right)^{-2} , \qquad (2\text{-}120)$$

indicating that C_D and z_0 are in one-to-one correspondence.

Realistic values of the bulk transfer coefficients are given, as a first approximation for a nearly neutral stratification, as follows: $C_D = (1\sim2)\times10^{-3}$ and

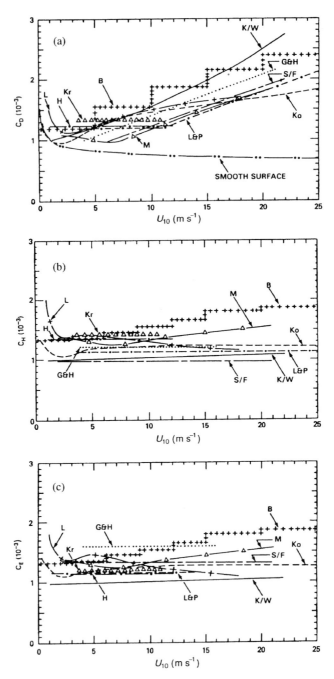

Fig. 2-10. Wind speed dependence of bulk transfer coefficients for momentum (drag coefficient, C_D), sensible heat (C_H), and water vapor (C_E), used in contemporary numerical models (after Blanc, 1985).

$C_H \fallingdotseq C_E = (1 \sim 1.5) \times 10^{-3}$. However, observed values scatter widely, and a number of researchers have proposed different formulas (see Fig. 2-10).

These coefficients are presumably dependent on wind speed at least. As seen from Fig. 2-10, C_D varies with U_{10} more acutely than C_H or C_E does. However, they should also depend on the sea-surface state, as expected from Eq. (2-32). In particular, the windsea is not simply an undulating sea surface, but something strongly coupled with the wind, as discussed in Section 2.4. Hence an attempt was made to express the dimensionless roughness parameter as a function of the ratio of the phase speed c_p of the representative windsea to U_{10} (*wave age*), or the ratio of u_* to c_p (Kusaba and Masuda, 1988),

$$\frac{g z_0}{u_*^2} = \gamma \left(\frac{\sigma_p u_*}{g} \right)^\varepsilon \qquad (2\text{-}121)$$

where γ and ε are both constants. In this case, if one particular series of field observational data is selected, or certain wind tunnel data is used, ε tends to be near 1, whereas if all kinds of observations are included, ε is estimated to be about -0.5. This discrepancy apparently comes from the presence of short-time scale fluctuation of wind (*gustiness*, change in directions, etc.), swells from different directions, currents, and other effects.

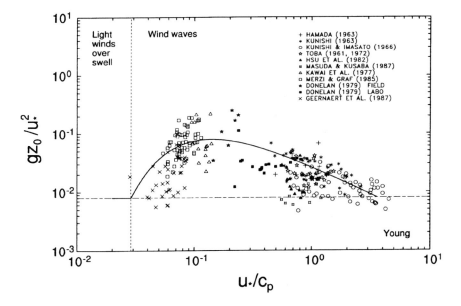

Fig. 2-11. A formula of the nondimensional roughness parameter $g z_0/u_*^2$ as a function of inverse wave age u_*/c_p proposed by scientists of an international working group SCOR 101, together with a data set of observed values (after Jones and Toba, 2001).

Replacing ε in Eq. (2-121) with 0 will give Charnock's formula (Charnock, 1955),

$$z_0 = \frac{\gamma u_*^2}{g}, \quad \gamma = 0.0185.$$
(2-122)

Figure 2-11 shows, together with a data set of observed values, a formula recently proposed by scientists of an international working group, of the *nondimensional roughness parameter* gz_0/u_*^2 as a function of u_*/c_p (Jones and Toba, 2001):

$$\frac{gz_0}{u_*^2} = \begin{cases} \left[0.03\left(\dfrac{u_*}{c_p}\right)^{-1} \exp\left[-0.14\left(\dfrac{u_*}{c_p}\right)^{-1}\right], & \sim \dfrac{1}{0.35} > \dfrac{u_*}{c_p} > \dfrac{1}{35} \\ 0.008, & \dfrac{u_*}{c_p} \le \dfrac{1}{35} \quad \text{(light winds over swell)}. \end{cases}$$

It is known that the spectra of fluctuating physical quantities in the atmospheric boundary layer have a region of very small energy in a range from 10 minutes to several hours. Thus the quantities used for the bulk formulas discussed above are assumed to be used for average values over approximately 10-minute long observations at the 10-m height. Accordingly it should be noted that, if, say, a monthly averaged wind speed is used for the bulk coefficient, the estimated values can be smaller by 30% (Hanawa and Toba, 1987).

2.5.3 Wave breaking, bubble entrainment, and generation of sea-water droplets

Since breaking of windsea at the sea surface is expected to affect the ocean–atmosphere interactions, especially at strong wind regions, its parameterization is very important.

To quantify wave breaking, photographing of the sea surface has been used to express the area percentage of whitecap coverage. Most researchers, however, treated it as a function of U_{10} only (e.g., Monahan and MacNiocaill, 1986; Wu, 1990).

In view of the situation that windsea is in local equilibrium with the wind as discussed in the present chapter, another parameter representing the windsea is better to be included in addition to the wind speed. Along this line, the following nondimensional *wave breaking parameter* was proposed (Toba and Koga, 1986):

$$R_B = \frac{u_*^2}{\nu \sigma_p}.$$
(2-123)

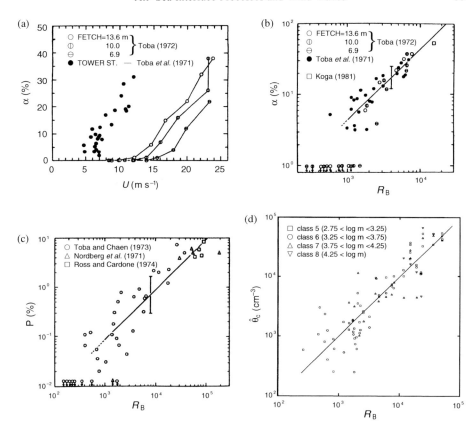

Fig. 2-12. Usefulness of breaking wave parameter, R_B of Eq. (2-123). (a): Percentage α of breaking crests among individual waves of windsea passing at a fixed point on the water surface, plotted against U_{10}, showing that α greatly depends on the fetch (after Toba and Koga, 1986). (b): Same data as (a) plotted against R_B of Eq. (2-123), demonstrating that all the data collapse along the line with an angle of 45° (after Toba and Koga, 1986). (c): The whitecap coverage percentage plotted against R_B (after Toba and Koga, 1986). (d): Sea-water droplet concentration observed near the sea surface plotted against R_B. The observed values are normalized for various size classes to come at the same level (after Iida et al., 1992).

This corresponds to a Reynolds number which contains $L_s = u_* T_s$ ($=2\pi u_*/\sigma_p$) as a representative length scale of the phenomenon in consideration, where σ_p is the peak angular frequency of the windsea part of the wave spectra. The L_s is interpreted as the length representing the distance over which a water particle at the surface is driven by the tangential stress of wind distributed along each wave phase of an individual wave, during the representative wave period T_s.

In Fig. 2-12(a), the percentage α of breaking crests among individual waves of windsea passing at a fixed point on the water surface is plotted against U_{10}. The results of short-fetch wind-wave tank experiments systematically differ from those obtained at a tower station in the sea. This shows that α greatly depends on the fetch (which indicates the degree of development of windsea). If α is plotted against R_B in Eq. (2-123), all the data collapse along a line with an angle of 45°, as shown in Fig. 2-12(b). Figure 2-12(c) shows the whitecap coverage percentage plotted against R_B; here also all the data sets converge along the 45°-line. It should be noted that wave breaking starts to occur at around $R_B = 10^3$. The sea-water droplet concentration observed near the sea surface also shows a similar distribution along the 45°-line as shown in Fig. 2-12(d). Thus R_B is considered as an important parameter describing macroscopic features of wave breaking, air entrainment and sea-water droplet (sea-salt particles) production. Zhao and Toba (2001) proposed a regression formula of the percentage whitecap coverage P with R_B,

$$P = 3.88 \times 10^{-5} R_B{}^{1.09},$$

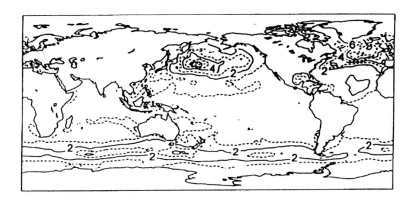

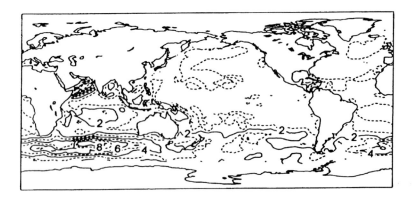

Fig. 2-13. An estimate of global distribution of whitecap coverage in January and July, from distribution of wind and windsea using R_B (after Toba *et al.*, 1999).

with a correlation coefficient of 0.88. They reported that other regression formulas of P with u_*, U_{10}, T_S and c_p/u_* had much smaller correlation coefficients for the same data set.

The effect of sea-water droplets on sea-surface evaporation is expected to be quite large when the wave height is several meters and the wind speed exceeds 20 m s^{-1} (Iida and Toba, 1999).

In recent years, the problem of wave breaking has attracted considerable attention in relation to the exchange process of carbon dioxide and other substances through the ocean–atmosphere boundary. Studies on the generation and distribution of bubbles into the sea and droplets (sea-salt particles) into the air have revived again. Acoustic devises good for measuring bubbles have also been developed, and how to quantify the generation and distribution of bubbles using such devices is being studied extensively (e.g., Thorpe, 1992).

Figure 2-13 shows an estimate of global distribution of whitecap coverage in January and July, from distribution of wind and waves using Eq. (2-123). It is interesting to note that there are some similarities between the areas of high whitecap coverage and the large-scale oceanic subduction regions. These areas are near Greenland and the western North Pacific in northern winter, and some part of the Southern Ocean in southern winter. This may induce further studies on the fate of CO_2 of the anthropogenic origin.

2.5.4 Gas exchange through the sea surface

Exchange of gas between the atmosphere and the ocean is an important subject in relation to the global warming problem. In the field of chemical engineering, it is assumed that, for convenience sake, there are two boundary films above and below the water surface, where molecular diffusion is predominant (see Fig. 2-14). Since CO_2 or O_2 has relatively small solubility for water, we can ignore the diffusion resistance of the air side, put $C_{GI} \sim C_{LI}$, and only consider the diffusion resistance of water. Then the flux F through the water surface is given by

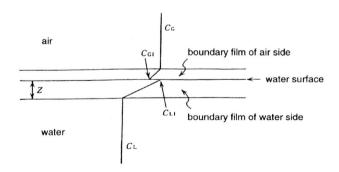

Fig. 2-14. Schematics of the double boundary film hypothesis concerning gas exchange at the air–water interface.

$$F = -D\frac{\partial C_L}{\partial z}\bigg|_{z=0}, \qquad (2\text{-}124)$$

where C_L and D are the concentration and the molecular diffusion coefficient of the gas, respectively, in water. If the gas exchange velocity k_L (m s^{-1}) is used, F can be written as:

$$F = k_L \Delta C, \quad \Delta C = C_{LI} - C_L. \qquad (2\text{-}125)$$

Or,

$$F = k_L S \Delta P \qquad (2\text{-}126)$$

where ΔP is the difference between the partial pressures of the gas in air and in water, and S the solubility of the gas. The value of k_L is to be determined empirically (see Fig. 2-15). The thickness $Z\,(=D/k_L)$ of the boundary film in water

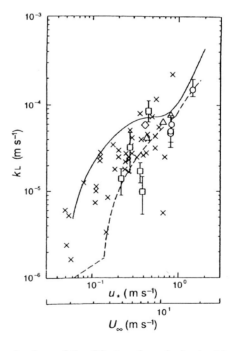

Fig. 2-15. Observed values of the CO_2 transfer velocity k_L. Most data points are measured at sea, the broken line is the relation by Liss and Merlivat (1986) determined from these points, and the solid line indicates observed values in a wind-wave tank (after Komori *et al.*, 1995).

(see Fig. 2-14) is roughly estimated to be about 200 μm based on the experimentally determined k_L.

It is to be noted that in the real situation, turbulent eddies reach around 200 μm, and there is no molecular diffusion film there. Also, the observed value of k_L varies widely as seen from Fig. 2-15. According to laboratory experiments, the exchange of momentum (e.g., Toba and Kunishi, 1970) and that of gas tend to vary in a similar way (Komori et al., 1995). This subject remains to be studied further.

2.6 WAVE FORECASTING

Wave forecasting is an important subject not only for applications to navigation and coastal ocean questions, but for the parametrization of sea surface processes discribed in the previous section, and also for the satellite remote sensing to be discussed in Chapter 7.

2.6.1 Energy equilibrium equation

In models for estimation of ocean waves, it is usual to assume that the variations in time t and space $x(x, y)$ of a two-dimensional spectrum $F(f, \theta)$ or $F(k)$ is controlled by the *energy balance equation*

$$\frac{\partial F}{\partial t} + c_g \cdot \nabla F = S_{net} , \quad \nabla \equiv \left(\frac{\partial}{\partial x}, \frac{\partial}{\partial y} \right), \tag{2-127}$$

where $c_g(f, \theta)$ is the group velocity of component waves in the spectrum, f the frequency and θ the direction. S_{net} on the right-hand side is called the *source function*. If the windsea is controlled by strongly nonlinear processes as discussed in Section 2.4, it is rather difficult to separate the source function into linear components, but it is traditionally divided into three processes. Namely,

$$S_{net} = S_{in} + S_{nl} + S_{ds} \tag{2-128}$$

where S_{in} is the energy input from the wind, S_{nl} is the redistribution of the energy among wave components by wave interactions due to the four-wave resonance expressed by Eq. (2-87), and S_{ds} is the energy dissipation by wave breaking. Consequently, the quantitative expressions of these source functions are the most crucial point for wave forecasting models. For now, certain observed values are usually used for S_{in}, and some simplified calculations are adopted for S_{nl}. The form of S_{ds} is tuned for the calculated wave growth to be in accordance with the empirical growth, for values of adopted S_{in} and S_{nl}. Figure 2-16 shows an example of these source functions relative to a one-dimensional spectral form of windsea, calculated by the RIAM method (Masuda, 1995), which will be discussed in the subsequent section.

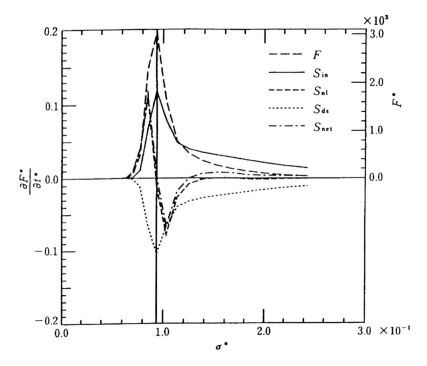

Fig. 2-16. An example of source functions relative to the one-dimensional spectral form of windsea, calculated by the RIAM method (after Masuda, 1995).

2.6.2 Recent wave models and further research subjects

In early years, the nonlinear interaction term in Eq. (2-128) was regarded as insignificant, and "decoupled propagation models" in which each component wave develops and decays independently, were mainly adopted.

Since the 1970s, models of wind-wave growth including nonlinear interactions have been developed. Hasselmann *et al.* (1976) proposed a model with parameters simplified from the results of multitudinous calculations based on a weakly nonlinear interaction theory. Also, based on the 3/2-power law (Eq. (2-101)) in another form which was expressed by the nondimensional energy E^* ($=gE/u_*^4$) related with the nondimensional angular frequency σ_p^* ($=u_*\sigma_p/g$):

$$E^* = 5.1 \times 10^{-2} \sigma_p^{*-3}, \qquad (2\text{-}129)$$

the TOHOKU wave model (Joseph *et al.*, 1981; Toba *et al.*, 1985a, b) was developed. It uses the following prediction equation (Toba, 1978), which traces wind-wave growth by only one parameter E^* as a function of the nondimensional time t^* ($=gt/u_*$):

$$\left.\begin{array}{l} \dfrac{D\!\left(E^{*2/3}\right)}{Dt^{*}} = 2.4 \times 10^{-4}\left\{1 - \mathrm{erf}\!\left(0.12 E^{*1/3}\right)\right\} \\[3mm] \mathrm{erf}(\zeta) = \dfrac{2}{\sqrt{\pi}} \displaystyle\int_{0}^{\xi} \exp\!\left(-\xi^{2}\right) d\xi \end{array}\right\}. \qquad (2\text{-}130)$$

These and other wave models, in which the windsea part and swell components in the spectrum are predicted with some mutual transfer, are thus called coupled hybrid models. The MRI-II model (Uji, 1984), which had been in use for ten or more years by the Japan Meteorological Agency (JMA) for routine wave forecasting, adopted Eq. (2-130) for developing windsea while integrating Eq. (2-127) with Eq. (2-128) for component waves; these models are called "coupled discrete models".

In Europe the WAve Modelling group led by Hasselmann (The WAMDI Group, 1988; Komen *et al.*, 1994) was active in developing a model for forecasting waves globally, as well as locally and in the shallow sea, while assimilating

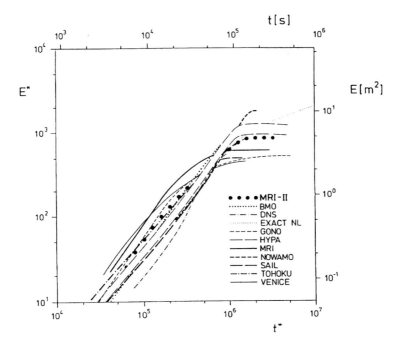

Fig. 2-17. The nondimensional energy E^{*} for a stationary and uniform wind plotted against the nondimensional wind duration t^{*} ($=gt/u_{*}$). MRI-II values (Uji, 1984) are added to the results of an international comparison experiment of wave models (The SWAMP Group, 1985). The upper and the right side dimensional scales indicate values for a case of U_{10} of 20 m s^{-1}.

satellite data in an atmospheric general circulation model. It adopts the calculation scheme of the nonlinear energy transfer term S_{nl} with much more freedom, and is called the "third generation ocean wave prediction model". Related to this category, in Japan Komatsu *et al.* (1993) developed the RIAM method, which retained an accuracy of the exact calculation of S_{nl} (Masuda, 1980) using a simplified high-speed algorithm. JMA now operates a newly developed MRI-III wave model for routine forcasting, and Japan Weather Association (JWA) has developed the JWA3G Model by Suzuki (1995), all being third generation models.

Empirically, the windsea growth in a simple condition is seen by the empirical formulas from Eqs. (2-96) to (2-100), and Eq. (2-130) corresponds to Eqs. (2-96) and (2-97) converted by use of the 3/2-power law Eq. (2-101) or (2-129). In Fig. 2-17, the nondimensional energy E^* for a stationary and uniform wind pattern is plotted against the nondimensional wind duration t^* ($=gt/u_*$) from the result of an international comparison experiment of wave models (The SWAMP Group, 1985) with an additional calculation using MRI-II. If the fetch is chosen in place of the duration, the result is a similar figure. The wave energy increases in proportion to fetch at initial stages, but it ceases to develop after the nondimensional duration or fetch reaches a certain value, toward a saturation stage.

Developing windsea turns into swells when the wind ceases. On the contrary, a swell becomes windsea when the wind builds up. Since the phenomenon of windsea is strongly nonlinear as described in Section 2.4, if the direction of the wind is not much different from that of a swell, the swell becomes windsea in that direction rather quickly. If the wind blows stronger in the direction considerably different from that of a swell, the situation will be much more complex. This cannot be treated purely theoretically, and although laboratory experiments and observational studies (e.g., Holtheijsen *et al.*, 1987) are being conducted, they have not yet provided sufficient results. In addition to the question of the wind stress as discussed in Section 2.5.2, the changing wind field over the complex wave field is still the most difficult subject for wave models.

Masuda (1995) suggested the existence of further energy transfer mechanisms that cannot be expressed by existing equations, since the process of wave development in a wind-wave tank experiment was not well reproduced by the RIAM method. This is apparently the question directly connected with the strong coupling between the windsea and the wind, and so also with the local equilibrium described in Section 2.4. Hence, the dynamics of wind waves including the particular forcing of the wind and wave breaking is still a forefront research topic in fluid mechanics.

REFERENCES

Bailey, R. J. *et al.* (1991): *J. Oceanogr. Soc. Japan*, **47**, 249–264.
Banner, M. L. and O. M. Phillips (1974): *J. Fluid Mech.*, **65**, 647–656.
Blanc, T. V. (1985): *J. Phys. Oceanogr.*, **15**, 650–669.
Brooke Benjamin, T. and J. E. Feir (1967): *J. Fluid Mech.*, **27**, 417–430.

Bye, J. A. T. (1988): *J. Mar. Res.*, **46**, 457–472.

Charnock, H. (1955): *Quart. J. Roy. Meteor. Soc.*, **81**, 639–640.

Crapper, G. D. (1957): *J. Fluid Mech.*, **2**, 532–540.

Csanady, G. T. (2001): *Air-Sea Interaction: Laws and Mechanisms*, Cambridge Univ. Press, 239 pp.

Ebuchi, N. *et al.* (1987): *Boundary Layer Meteor.*, **39**, 133–151.

Ebuchi, N. *et al.* (1992): *J. Oceanogr.*, **48**, 77–92.

Hanawa, K. and Y. Toba (1987): *Ocean-Air Int.*, **1**, 79–93.

Hasselmann, K. *et al.* (1973): *Deutsch. Hydrogr. Z.*, **12**, 1–95.

Hasselmann, K. *et al.* (1976): *J. Phys. Oceanogr.*, **6**, 200–228.

Hatori, M. (1984): *J. Oceanogr. Soc. Japan*, **40**, 12–18.

Holtheijsen, L. H. *et al.* (1987): *J. Phys. Oceanogr.*, **17**, 845–853.

Huang, N. E. (1998): *Proc. Roy. Soc. Lond.*, **A454**, 903–995.

Huang, N. E. *et al.* (1990): *The Sea—Ocean Science Engineering*, Vol. 9, Part A, ed. by B. LeMehaute and D. M. Hanes, Wiley, pp. 197–237.

Iida, N. and Y. Toba (1999): *The Wind-Driven Air-Sea Interface,* ed. by M. L. Banner, School of Math., Univ. of NSW.

Iida, N. *et al.* (1992): *J. Oceanogr.*, **48**, 439–460.

Jähne, B. and K. S. Riemer (1990): *J. Geophys. Res.*, **95**, 11531–11546.

Jones, I. S. F. and Y. Toba (eds.) (2001): *Wind Stress over the Ocean*, Cambridge Univ. Press, 307 pp.

Joseph, P. S. *et al.* (1981): *Tohoku Geophys. J.*, **28**, 27–45.

Kawai, S. (1979): *J. Fluid Mech.*, **93**, 661–703.

Kawai, S. *et al.* (1977): *J. Oceanogr. Soc. Japan*, **33**, 137–150.

Kawamura, H. and Y. Toba (1988): *J. Fluid. Mech.*, **197**, 105–138.

Kitaigorodskii, S. A. (1961): *Bull. Acad. Nauk SSSR Geophys. Ser.*, 105–117.

Koga, M. (1981): *Tellus*, **33**, 552–563.

Komatsu, Y. *et al.* (1993): *Report of Research Institute for Applied Mechanics, Kyushu University*, **75**, 121–146 (in Japanese).

Komen *et al.* (1994): *Dynamics and Modelling of Ocean Waves*, Cambridge Univ. Press, 532 pp.

Komori, S. *et al.* (1995): *Biogeochemical Processes and Ocean Flux in the Western Pacific*, ed. by H. Sakai and Y. Nozaki, Terra Sci. Pub. Co., Tokyo, pp. 69–81.

Kusaba, T. and A. Masuda (1988): *J. Oceanogr. Soc. Japan*, **44**, 200–214.

Liss, P. S. and L. Merlivat (1986): *The Role of Air-Sea Exchange in Geochemical Cycling*, ed. by P. Buat-Menard, D. Reidel, pp. 113–127.

Longuet-Higgins, M. S. (1952): *J. Mar. Res.*, **11**, 245–266.

Longuet-Higgins, M. S. (1972): *Proc. Roy. Soc. Lond.*, **A331**, 445–456.

Longuet-Higgins, M. S. (1975): *Proc. Roy. Soc. Lond.*, **A342**, 157–174.

Longuet-Higgins, M. S. and R. W. Stewart (1964): *Deep-Sea Res.*, **11**, 529–562.

Longuet-Higgins, M. S. *et al.* (1961): *Ocean Wave Spectra,* Prentice-Hall, Englewood Cliffs, pp. 111–132.

Masuda, A. (1980): *J. Phys. Oceanogr.*, **10**, 2082–2093.

Masuda, A. (1995): *Nagare*, **14**, 198–208 (in Japanese).

Masuda, A. *et al.* (1979): *J. Fluid Mech.*, **92**, 717–730.

McLean, J. W. (1982): *J. Fluid Mech.*, **114**, 331–341.

Miles, J. W. (1957): *J. Fluid Mech.*, **3**, 185–204.

Mitsuyasu, H. (1971): *Coast. Eng. Japan*, **14**, 1–14.

Mitsuyasu, H. and T. Honda (1974): *J. Oceanogr. Soc. Japan*, **30**, 185–198.

Mitsuyasu, H. *et al.* (1975): *J. Phys. Oceanogr.*, **5**, 750–760.

Mitsuyasu, H. *et al.* (1979): *J. Fluid Mech.*, **92**, 731–749.

Monahan, E. C. and G. MacNiocaill (eds.) (1986): *Oceanic Whitecaps and Their Role in Air-Sea Exchange Processes*, D. Reidel, Dordrecht, 294 pp.

Monin, A. S. and A. M. Obukhov (1954): *Trudy Geofiz. Inst. Acad. Nauk SSSR*, **24**, 163–187.

Penrose, R. (1989): *The Emperor's New Mind*, Oxford Univ. Press, 466 pp.

Phillips, O. M. (1957): *J. Fluid Mech.*, **2**, 417–445.

Phillips, O. M. (1977): *The Dynamics of the Upper Ocean*, 2nd ed., Cambridge Univ. Press, 336 pp.
Phillips, O. M. (1985): *J. Fluid Mech.*, **156**, 505–531.
Pierson, W. J., Jr. and L. Moskowitz (1964): *J. Geophys. Res.*, **69**, 5181–5190.
Prandtl (1932): *Beitrage zur Physik der freien Atmos.*, **19**, 188–202.
Schwartz, L. W. (1974): *J. Fluid Mech.*, **62**, 553–578.
Stokes, G. G. (1847): *Trans. Camb. Phil. Soc.*, **8**, 441–455.
Suzuki, Y. (1995): Development and application of a global ocean wave prediction model including nonlinear interactions and dissipation. Ph.D. Thesis, Graduate School of Science, University of Tokyo, 182 pp.
Sverdrup, H. U. and W. H. Munk (1947): *Wind, Sea and Swell—Theory of Relations for Forecasting*, U.S. Navy Hydrogr. Office. Wash., Pub. No. 601, 44 pp.
The SWAMP Group (1985): *Ocean Wave Modeling*, Plenum, 256 pp.
The WAMDI Group (1988): *J. Phys. Oceanogr.*, **18**, 1775–1810.
Thorpe, S. A. (1992): *Q. J. R. Meteorol. Soc.*, **118**, 1–33.
Toba, Y. (1972): *J. Oceanogr. Soc. Japan*, **28**, 109–120.
Toba, Y. (1973): *J. Oceanogr. Soc. Japan*, **29**, 209–220.
Toba, Y. (1978): *J. Phys. Oceanogr.*, **8**, 494–507.
Toba, Y. (1988): *Fluid Dyn. Res.*, **2**, 263–279.
Toba, Y. (1998): *Nonlinear Ocean Waves*, ed. by W. Perrie, Comp. Mech. Pub., pp. 1–59.
Toba, Y. and Y. Kawamura (1996): *J. Oceanogr.*, **52**, 409–419.
Toba, Y. and M. Koga (1986): *Oceanic Whitecaps*, D. Reidel, pp. 37–47.
Toba, Y. and H. Kunishi (1970): *J. Oceanogr. Soc. Japan*, **26**, 71–80.
Toba, Y. *et al.* (1985a): *Ocean Wave Modeling*, ed. by The SWAMP Group, Plenum, pp. 201–210.
Toba, Y. *et al.* (1985b): *The Ocean Surface*, ed. by Y. Toba and H. Mitsuyasu, D. Reidel, pp. 227–232.
Toba, Y. *et al.* (1988): *J. Phys. Oceanogr.*, **18**, 1231–1240.
Toba, Y. *et al.* (1999): *The Wind-Driven Air-Sea Interface*, ed. by M. L. Banner, School of Math., Univ. of NSW.
Tokuda, M. and Y. Toba (1982): *J. Oceanogr. Soc. Japan*, **38**, 8–14.
Tulin, M. P. (1996): *Waves and Nonlinear Processes in Hydrodynamics*, ed. by J. Grue *et al.*, Kluwer, pp. 177–190.
Uji, T. (1984): *J. Oceanogr. Soc. Japan*, **40**, 303–313.
Wilson, B. W. (1965): *Deutsch. Hydrogr. Z.*, **18**, 114–130.
Wu, J. (1990): *J. Geophys. Res.*, **95**, 18269–18279.
Yoshikawa, I. *et al.* (1988): *J. Oceanogr. Soc. Japan*, **44**, 143–156.
Zakharov, V. E. and N. N. Filonenko (1966): *Doklady Akademii Nauk SSSR*, **170**, 1291–1295.
Zhao, D. and Y. Toba (2001): *J. Oceanogr.*, **57**, 603–616.

I left Caen, where I was living, to go on a geologic excursion ... The incidents of the travel made me forget my mathematical work. ... we entered an omnibus to go to some place or other. At the moment when I put my foot on the step, the idea came to me, without anything in my former thoughts seeming to have paved the way for it, that the transformations I had used to define the Fuchsian functions were identical with those of non-Euclidean geometry. I did not verify the idea; I should not have had time, ... but I felt a perfect certainty. On my return to Caen, ... I verified the result at my leisure.

—Henri Poincaré (after Penrose, 1989)

Chapter 3

Surface Mixed Layer in the Ocean and Water Mass Analysis

Kimio HANAWA and Toshio SUGA

Ocean–Atmosphere Interactions, Ed. Y. Toba, pp. 63–109.
© by TERRAPUB / Kluwer, 2003.

Surface Mixed Layer in the Ocean and Water Mass Analysis

Kimio HANAWA[1] and Toshio SUGA[2]

The river is within us, the sea is all about us.

—T. S. Eliot

3.1 SURFACE MIXED LAYER IN THE OCEAN

The surface mixed layer in the ocean is the layer below the ocean surface where its water temperature and salinity, and in particular its density, are uniform in the vertical direction. Stratification of the ocean surface is formed by the incoming radiation energy from the sun, while turbulent mixing tends to destroy it. The two processes govern development or destruction of the mixed layer.

Among the time scales of solar radiation, daily and yearly (seasonal variation) periods are most pronounced. The mixed layer varies corresponding to the two time scales. The daily mixed layer is one above the daily thermocline, being formed and disappearing on the time scale of one day. The seasonal mixed layer above the seasonal thermocline deepens or shallows on the time scale of one year. Generally the term "surface mixed layer" implies the latter, unless otherwise stated.

3.1.1 Difference between atmospheric and oceanic mixed layers

In the lower part of the atmosphere near the ground, a mixed layer is also formed. Microscopically the physical process forming the mixed layer is the same as the ocean, but macroscopically it is different from the oceanic mixed layer. The mixed layer in the atmosphere is continually mixing, so that it would be best called a "mixing layer." That is, the atmospheric mixed layer is always turbulent to a certain degree. The atmosphere absorbs only a fraction of radiation from the sun and the thermal energy warms up the ground and ocean surface first. The heat in the ground and the ocean warms the atmosphere in the form of sensible and latent heats and upward long-wave radiation. Heated and expanded air mass in the lower part of the atmosphere rises and mixes with air mass in the upper atmosphere, destroying stratification. As a result in the daytime in summer when

---

[1]Sections 3.1 and 3.2
[2]Sections 3.3 and 3.4

solar radiation is active, heating of the ground can be so great that a mixed layer of uniform potential temperature is formed as high as 500–2000 meters. In the atmosphere only daily mixed layers exist and no seasonal mixed layer is formed. This is because the thermal inertia of air is extremely small. That is, the time scale of heating and cooling the atmosphere is shorter than that of the ocean.

On the other hand, stratification of the oceanic mixed layer is destroyed by a mechanical mixing due to wind stress and by a convective mixing due to the negative buoyancy force, namely the mixing caused by a high-density fluid formed near the sea surface by emitting heat and evaporation from the surface. As a result, a layer with almost uniform physical properties such as temperature, salinity, etc., that is, a mixed layer, is formed from the sea surface to a certain depth. The turbulence caused by wind stress is called forced convection, and the corresponding mixing is called mechanical mixing. The turbulence caused by heat emission and fresh water is called natural convection or thermal convection.

Seawater has much greater thermal inertia than air. Consequently, a small amount of incoming or outgoing heat will not greatly alter the structure of ocean stratification. Thus, the turbulence in the oceanic mixed layer is not very strong: once a certain state is created, it tends to last for a long time. The mixed layer in the ocean is not only mixing but also mixed. This is the reason why the seasonal mixed layer exists in addition to the daily mixed layer.

3.1.2 Meaning of researches on the oceanic mixed layer

We will consider the importance of the role of the mixed layer in the ocean. The mixed layer plays an important role in air–sea interactions. It is also important in the ocean itself in forming surface water masses.

a) Ocean–atmosphere interactions and oceanic mixed layer

Most of the solar energy that reaches the sea surface is absorbed in the upper layer of 100 meters in the ocean. The ocean distributes the heat vertically, stores and carries it to other areas by advection and releases it to the atmosphere. The state variable to determine the amount of the release from the ocean is determined by sea surface temperature. The sea surface temperature is given as a stationary boundary condition in a numerical model for short-term weather forecasting. Sea surface temperature is assumed to vary very little for a short period of time, although heat comes in and out through the sea surface.

There are various spatial and time scales in the distribution of sea surface temperature and its variations. Some examples in the North Pacific will be discussed in the following chapter. Long term variations of sea surface temperature on the large scale are consistent with large-scale variations of atmospheric circulation fields. However, how such large-scale anomalies of sea surface temperatures that last from a few months to several years come into existence, or how they influence the atmospheric circulation model, is not yet fully known. Since on this scale the ocean and the atmosphere are a united system with interactions that has a feedback circuit in it, an ocean–atmosphere-coupled numerical model will be most valid to elucidate the process. In such a model, the sea surface temperature, which is a given stationary lower boundary condition for the short-term forecasting model, is now a variable itself to be predicted. In order to forecast variations of sea surface temperature correctly, a model to describe the dynamics of the oceanic mixed layer or the mixed layer model is necessary.

b) Water mass formation in the surface layer and the mixed layer

The mixed layer process in the ocean surface layer is also a process for forming water masses. For example, off the Kuroshio, the western boundary current of the subtropical circulation in the North Pacific flowing near Japan, there exists subtropical mode water, which has the largest volume in the surface layer of the subtropical circulation systems. This water mass is distributed between the seasonal thermocline and the permanent thermocline from the western to the central regions of the subtropical circulation system with a thickness of up to a few hundred meters. This subtropical mode water is formed within a mixed layer with a thickness of 300–400 meters as a result of heat loss from the sea surface due to the monsoon outbreak over the sea in winter. The uniform water mass thus created is bounded by a new mixed layer made by heating from solar radiation in summer and is left in the subsurface layer.

It has recently been shown that central mode water is another water mass in the North Pacific that is also formed by the mixed layer process. This mode water is presumably made in winter, near the International Date Line, between the Kuroshio main stream and a Kuroshio branch. The North Pacific subtropical mode water and the North Pacific Central Mode Water will be discussed in detail in the next section. In the Labrador Sea, the Norwegian Sea and over the continental shelf in the Antarctic Ocean, deep water and bottom water are formed. Near the Antarctica a denser water at the surface sinks into the bottom like a chimney. The formation of this chimney can be called a mixed layer process in a broad sense.

The water mass formed in the surface mixed layer is sheltered from the atmosphere by a new mixed layer produced in summer. The mass in the surface layer is carried by the circulation field, which is created by a larger-scale dynamics. Then it is pushed downward by a vertical flow due to Ekman convergence (subduction), or to a lower layer and sometimes sinks to the bottom if a denser mass is formed. The water mass produced in the mixed layer rarely stays in the original area; it moves to a distant place.

In the process of winter mixed layer development, gases are actively exchanged between air and sea. For instance, subtropical mode water is saturated with oxygen dissolved into the seawater from the atmosphere. The dissolved gases are also transported to a distant place with the movement of the water mass created in the mixed layer. Carbon dioxide is an important gas that is exchanged between the ocean and the atmosphere. Carbon dioxide in the air is regarded as one of the elements to have the greatest impact on climate, and its artificial increase is causing a problem, that is, global warming. It is absorbed in the mid-to high-latitude regions of the ocean while it is emitted from low-latitude regions. Whether or not the ocean absorbs carbon dioxide as a total, or is only a buffer to smooth out variations of carbon dioxide in the atmosphere, remains to be studied. To understand the process, it is essential to study the mixed layer and circulation in the ocean.

From another viewpoint, it can be said that the oceanic mixed layer process provides an initial condition to water particles that circulate three-dimensionally in the ocean.

3.2 MODELS OF THE OCEANIC MIXED LAYER

3.2.1 Mixed layer models

While the role of the ocean with respect to climate change is understood, it is necessary that we also understand the sea surface temperature distribution and its formation process. Sea surface temperature can be macroscopically regarded as the temperature of the mixed layer. As mentioned earlier, it is used as a boundary condition at the bottom of the atmosphere in short-term weather forecasting models, while it is treated as a variable to be forecast in intermediate-to long-term weather forecasting models in which both the atmosphere and the ocean are coupled into one system.

It is necessary to use primitive equations to involve oceanic mixed layer processes in a numerical model for studying general circulation, etc. For dealing with the surface layer, a number of methods have been adopted. One is that the vertical diffusion coefficient is assumed to have an appropriate value in all layers, but if a heavier fluid is formed in the surface layer because of cooling and evaporation at the sea surface, the fluid is merged vertically so that the stratification becomes stable. This is called the convective adjustment method. The vertical diffusion coefficient in the merged fluid is infinite at that time. In this case, no turbulence is assumed to occur even if the wind stress is exerted.

In the second method the vertical diffusion coefficient is assumed to depend on the condition of stratification and flows. The Richardson number is used as a function to express the field. It is a ratio of the degree of stabilizing stratification and the degree of occurrence of turbulence and mixing due to velocity shear. Consequently, the vertical diffusion coefficient becomes a function of time and position. It is, in general, large at the surface and decreases as the depth increases.

In the third method, turbulent energy that causes mixing is explicitly treated at least near the surface. This is what is called the turbulent model of the mixed layer. Because its original equation of motion has nonlinear terms, the turbulent energy equation inevitably contains third-order products of velocity, etc. Hence, it is necessary to parameterize these at a certain level by the averaged gradients of temporally averaged variables and others. There are several levels depending on how accurately they are to be expressed. Closing equations is called the closure problem (Mellor and Yamada, 1974). This third method is recently becoming more popular for models to simulate general circulation in the basin-scale ocean.

The development of a mixed layer model in response to atmospheric disturbances was studied extensively from the late 1960s through the 1970s. It was a study of the "oceanic mixed layer model," that is, a model that is capable of expressing variations of the mixed layer responding to wind stress and the buoyancy force. Variations of the mixed layer are the temporal variations of its water temperature, salinity, current velocity (or their vertical distributions) and thickness. Those models studied in early years were all horizontally uniform and the physical process was one-dimensional.

The mixed layer model can be classified into two categories. One is the so-called "bulk model" or "integral model," while the other is the so-called "turbulence model" described earlier, or "continuous model." For representative samples of the two models, see Niiler (1975) and Mellor and Durbin (1975), respectively. See also review papers of Niiler and Kraus (1975) and Garwood (1979).

In the following section, physical processes to cause the mixed layer to vary will be reviewed by dealing with a bulk model.

3.2.2 Physical processes in the oceanic surface mixed layer

From the viewpoint of the bulk model, we will derive equations to express temporal variations of the water temperature and the thickness of the mixed layer, classify the processes related to the mixed layer variations, and summarize them (Hanawa and Toba, 1981).

a) Equations governing the temperature and the thickness of the mixed layer

As the local structure of the mixed layer, we assume a "slab model" of which the vertical structure is illustrated in Fig. 3-1. Since the velocity is assumed to be constant within the mixed layer, the layer behaves as if it were a slab. This slab

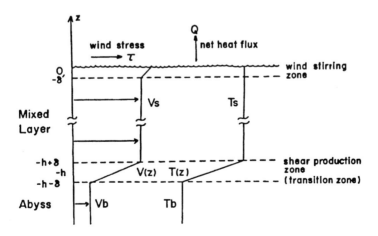

Fig. 3-1. Schematic picture of the vertical structure of the mixed layer model (Hanawa and Toba, 1981).

model is the most basic one of bulk models. Physical quantities considered here are assumed to be functions of time and place. The mixed layer is assumed to have a thickness of $h(x, y, t)$ and vertical homogeneous temperature $T_s(x, y, t)$. The horizontal velocity is also assumed to be vertically uniform except for the wind-stirring layer of a thickness δ. Below the mixed layer we assume a transition layer with a thickness of 2δ, where both temperature and horizontal velocity decrease rather abruptly to match the corresponding values in the lower layer, T_b and v_b, respectively. This transition layer is regarded as the entrainment layer where more turbulent fluid in the mixed layer entrains less turbulent fluid from the lower layer.

When water temperature and current velocity are expressed by the sums of the time-averages and fluctuations, the equation for the mean temperature of incompressible seawater is

$$\frac{\partial}{\partial t}T = -\frac{q}{\rho_0 C_p} - \nabla \bullet vT - \frac{\partial}{\partial z}wT - \nabla \bullet \overline{v'T'} - \frac{\partial}{\partial z}\overline{w'T'} \qquad (3\text{-}1)$$

where ρ_0 is the reference density of water, and C_p the specific heat at a constant pressure. The over-bar that indicates the mean value is omitted except for the correlation quantities. q is the heat source from the short wave solar radiation according to the following expression proposed by Denman (1973).

$$q = -\gamma Q_s e^{\gamma z} \qquad (3\text{-}2)$$

where Q_s represents the incident net solar radiation at the sea surface, and γ the extinction coefficient, which is in a range of 10–30/m.

First we express the vertical flux of heat at the bottom of the mixed layer by the entrainment velocity w_e, which can be defined as the moving velocity of the boundary surface not accounted for by the vertical current velocity. That is, the location of the bottom of the mixed layer is specified by

$$\frac{d}{dt}(-h) = w\big|_{-h} + w_e \ . \qquad (3\text{-}3)$$

If we apply this formula formally to the top and bottom of the transition layer and we use an integral form of Eq. (3-1), then we can obtain the turbulent heat flux specified by the entrainment velocity

$$\left(T_s - T_b\right)w_e = \overline{v'T'_+} \bullet \nabla h + \overline{w'T'_f} \qquad (3\text{-}4)$$

where the subscript + indicates the value at the top of the transition layer. When the boundary is tilted, the flux due to horizontal diffusion is also expressed by the entrainment term.

We put the heat flux at the surface as

$$-\overline{w'T'}\Big|_0 = -\frac{Q_B + Q_E + Q_H}{\rho_0 C_p} \qquad (3\text{-}5)$$

where Q_B, Q_E and Q_H are heat fluxes by long-wave back radiation, latent heat and sensible heat, respectively, and we assume that the net shortwave radiation is absorbed into the mixed layer, namely

$$-\int_{-h+\delta}^{0} \frac{q}{\rho_0 C_p} dz \sim \frac{Q_s}{\rho_0 C_p}. \qquad (3\text{-}6)$$

Then the integration of Eq. (3-1) will lead to the following equation that describes temporal variations of seawater temperature in the mixed layer,

$$\frac{\partial}{\partial t} T_s = \frac{1}{h}\left\{ -\frac{Q}{\rho_0 C_p} - (T_s - T_b)w_e - h\boldsymbol{v}_s \bullet \nabla T_s - \nabla \bullet h\overline{\boldsymbol{v}_s' T_s'} \right\} \qquad (3\text{-}7)$$

where $Q \equiv Q_S + Q_B + Q_E + Q_H$.
From Eq. (3-3) we can obtain the equation for the thickness of the mixed layer:

$$\frac{\partial}{\partial t} h = w_e - h\nabla \bullet \boldsymbol{v}_s - \boldsymbol{v}_s \bullet \nabla h. \qquad (3\text{-}8)$$

If we rewrite Eqs. (3-6) and (3-7),

$$\frac{\partial}{\partial t} T_s = \frac{1}{h}\left\{ -\frac{Q}{\rho_0 C_p} - (T_s - T_b)w_e - h\boldsymbol{v}_s \bullet \nabla T_s - \nabla \bullet h\overline{\boldsymbol{v}_s' T_s'} \right\} \qquad (3\text{-}9)$$

One–dimensional process   Three–dimensional process

$$\frac{\partial}{\partial t} h = -w_e - h\nabla \bullet \boldsymbol{v}_s - \boldsymbol{v}_s \bullet \nabla h. \qquad (3\text{-}10)$$

One–dimensional process   Three–dimensional process

If the process of the mixed-layer variation is one-dimensional, the mixed-layer temperature is determined only by the net heat exchange through the seas surface and the entrainment of seawater from the lower layer, and the thickness of the mixed layer increases only through the entrainment process.

Three-dimensional processes should occur because of spatial non-homogeneity of temperature, velocity and thickness. Temperature variations consist of horizontal currents and horizontal mixing, both of which can be classified according to the temporal scales of the phenomena under consideration.

b) Classification of physical processes in the mixed layer variation

We will classify the physical processes in the mixed layer variation based on Eq. (3-10) (see Fig. 3-2 for the schematic diagram of classification).

The first term on the right-hand side is the entrainment term. As mentioned earlier the entrainment is the phenomenon in which one layer with stronger turbulence takes in the fluid of the other layer with weaker turbulence through the boundary between the two layers. There are two energy sources that induce turbulent motion and entrainment at the bottom of the mixed layer. One is the inflow of potential energy, the other the inflow of kinetic energy. The motion caused by the former is called thermal (free) convection. The latter can be divided into two categories: one goes into "turbulence," the other into the "mean current." The kinetic energy into "turbulence" is further classified into two

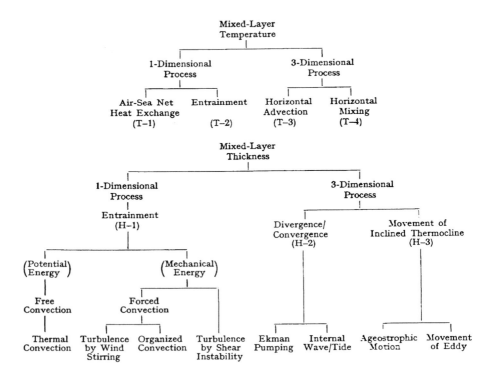

Fig. 3-2. Classification of the mixed-layer variations. The number in the parentheses corresponds to the term in Eqs. (3-9) and (3-10) (Hanawa and Toba, 1981).

different sources. One is the surface-originated turbulence directly associated with windsea (see Chapter 2), and the other Langmuir circulation, where the vorticity of the shear flow is more or less organized in the direction of the wind (e.g., Craik and Leibovich, 1976). These two types of "turbulence" can be called "forced convection" in contrast to free convection. The energy inflow into the mean current will include the Ekman current and inertial oscillation, which produce turbulence and entrainment due to the shear flow at the bottom of the mixed layer with density stratification. In short, we can regard the entrainment term as a result of the four processes occurring in the mixed layer.

The second term of Eq. (3-10) represents the effect of the non-homogeneity of current velocity, that is, convergence or divergence. It is caused by Ekman divergence or convergence due to irregularity of the wind field and the vertical displacement of thermocline due to internal waves and semidiurnal and diurnal tides.

The third term indicates the effect of movement of a tilted thermocline. This includes the movement of eddies in the presence of large-scale mean flow and ageostrophic components of the fluctuating thermocline.

Although we have provided an across-the-board classification above, the essential processes for the mixed-layer variations vary depending on spatial and temporal scales, ocean areas and seasons. For instance, surface-originated turbulence is a dominant process for the spatial and temporal scales matching to atmospheric disturbances. Free convection is important for a shorter-time scale of one day, and it also plays an essential role for the development of the mixed layer in cooling seasons. In the developing process of convergence or divergence of horizontal velocities, semidiurnal and diurnal internal tides can be averaged out over one period or longer, but the change of the thickness due to large-scale wind field variations should be studied, taking into account the scale of the wind field variability.

c) Relationship between the Ekman layer and mixed layer

The surface layer, when the wind stress is applied on the sea surface, forms the Ekman layer (Ekman, 1905). The current velocity at the surface in a homogeneous ocean with a constant eddy viscosity coefficient is deviated by 45 degrees to the right of the wind vector in the Northern Hemisphere. The current velocity vector in the Ekman layer forms the Ekman spiral, rotating clockwise as it goes down with its magnitude decreasing logarithmically. The depth of the Ekman layer is proportional to the square root of the eddy viscosity coefficient divided by the Coriolis parameter. For the eddy viscosity coefficient of 100 cm^2/sec in mid latitudes, the depth of the Ekman layer will be 10–20 m. What is the relationship between the mixed layer and the Ekman layer?

This relationship has not been fully clarified yet. First of all, the condition of the constant eddy viscosity coefficient is not realistic. Recent studies (e.g., Craig and Banner, 1994; Toba and Kawamura, 1995) have revealed that a turbulent layer directly associated with wind waves exists near the sea surface in addition to a Langmuir circulation. Consequently, the direction of the current velocity at the surface is not 45°. Also, if the depth of the mixed layer is smaller than that of the Ekman layer, the current velocity will decrease at this depth with a sudden change of its direction. This will bring a larger velocity shear and strong turbulence is expected to occur, which will induce further mixing. As a result, the mixed layer will be developed. In this case, the development of the Ekman layer coincides with the development of the mixed layer.

It is known that the distribution of the Ekman current velocity becomes unstable beyond a certain critical Reynolds number (Tatsumi and Goto, 1976). This means, that the developed Ekman layer cannot exist without becoming unstable. Measurements of the current velocities in the surface layer show that they are distributed more or less like a slab as assumed in the bulk layer model. There are very few studies on the instability of the Ekman layer in the real ocean. The reason why the Ekman spiral has rarely been observed may be due to the instability. The Ekman layer should be more studied both theoretically and observationally. The problem of the response of the Ekman layer to wind stress is an old but modern one.

d) Bulk model

Since Kraus and Turner (1967) first introduced a bulk model, numerous types of bulk models have been developed. A key point in the bulk model is how to parameterize the entrainment velocity. Two possible causes of how the turbulent energy occurs were first in consideration. One was the effect of wind stirring and the buoyancy force. The other was the velocity shear of the mean flow in the bottom part of the mixing layer. These two effects are later combined into one systematic model. Nevertheless, the formularization of the entrainment velocity will require several constants provided from outside (or artificially) and they are selected to match observations. The mathematical procedure of formulating the entrainment velocity in bulk models is overviewed by Masuda (1981).

The above-mentioned bulk models mainly deal with deepening of the mixing layer by kinetic energy. On the other hand, Woods (1980) maintained that deepening by the buoyancy force is essential. He precisely calculated the solar radiation absorbed in the ocean and discussed the thermal compensation depth. He showed that deepening of the mixed layer is caused by thermal convection except for short daytimes in summer. At the same time he pointed out that physical processes of daily time scales should be included in a seasonal model, and seasonal variation processes in an interannual model. De Szoeke (1980) made an attempt to extend the bulk model to a three-dimensional case, assuming that advection, convergence and divergence due to spatial variations of winds affect the distributions of temperature and depth of the mixing layer. Price (1981) developed a three-layer three-dimensional model to simulate the response of an oceanic upper layer to the passing of a hurricane.

The bulk models were extensively studied in the 1970s. Recent developments in computers have made it possible to introduce turbulent models without restriction, and so the role of bulk models has been reduced to analytical studies, not being of great use for numerical models. Nonetheless, the bulk model is still a very illustrative tool to understand the physical processes.

3.3 METHODS OF ANALYZING SURFACE WATER MASSES

Most water masses in the surface layer are formed by mixed layer processes (Section 3.1). What kind of water masses are formed? What are their formation rate? How do they circulate in the ocean? These problems have been studied as main topics of oceanography for a long time. Recent progress in the methods of oceanographic observations and analyses of water masses has made it possible to deal with the problems in a more quantitative manner. In this section, the concept of water masses will be reviewed first and then the main methods of analyzing them will be described.

3.3.1 Definition of water masses

A water mass is a fundamental term in oceanography. In some cases the word is used in a narrow sense based on the clear definition. In others it is used rather vaguely. For the terminology and its history in detail, see Tsuchiya (1970) or Masuzawa and Hasunuma (1977). The definitions in both narrow and broad senses will be described below, but this chapter will adopt the broad sense of the word.

A T–S diagram is a graph showing the relationship between temperature (T) and salinity (S) observed together as a function of depth (or pressure) in a vertical column of ocean water. Usually T is plotted in the ordinate against S in the abscissa. It is known that within a considerably large geographical region of the ocean, T–S curves are similar, but those from different regions show distinctly different curves. Therefore, T–S curves can be used to distinguish layer structures of seawater. In a narrow sense a water mass is defined as a body of water that has approximately the same T–S curves.

As an example, an averaged T–S curve using all the data of temperature and salinity stored in a data base before the end of the 1970s is shown in Fig. 3-3 (Emery and Dewar, 1982). The North Pacific and the North Atlantic were divided into 20 regions respectively and a T–S curve was plotted for each region. Figure 3-3 shows the averaged T–S curve for the central North Pacific region with standard deviations. From 8°C to 18°C the relationship between T and S is linear with very little deviation. This portion of the T–S curve corresponds to the West North Pacific Central Water defined by Sverdrup *et al.* (1942).

As discussed above, a curve or a narrow belt of curves on a T–S diagram represents a water mass in a narrow sense of the word. A point on a T–S diagram, on the other hand, is called a water type. According to Masuzawa and Hasunuma (1977), the concept of water types comes from the idea that if a T–S curve is a collection of several lines, seawater on each line is formed by the mixing of the seawater on both ends of the line (see Section 3.3.3(b)). In practice the seawater which is approximately represented by a point on a T–S diagram is often

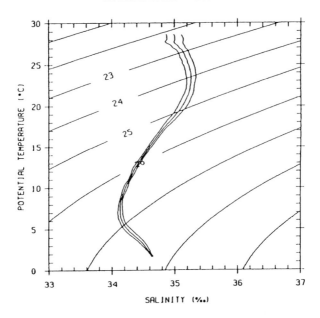

Fig. 3-3. A mean temperature–salinity curve averaged over the central North Pacific. The center mean line is bracketed by the standard deviations. Background contours are potential density, σ_θ (Emery and Dewar, 1982).

called a water type (Tsuchiya, 1970), but the water type in such usage is included in the water mass in a broad sense that will be discussed subsequently.

Although such a strict definition of a water mass based on a T–S curve exists, in general the word "water mass" presently in use has a more vague meaning. Here bodies of water that can be identified by particular combinations of physical and chemical characteristics are called water masses (Open University, 1989). A body of water that has nearly completely uniform characteristics, namely, a water type represented by a single point on a T–S curve, can be regarded as a special case of the water mass, and thus the unified term water mass will be used.

If we adopt the above definition on the water mass, we can define as many water masses as we like. However, in practice, a proper name and attention to its formation and circulation is given only to the water mass corresponding to an extremal value (e.g., salinity minimum, oxygen maximum) in the vertical distribution of seawater characteristics, or the water mass whose volume is relatively large in a certain region, that is, the water mass comprising the mode as used in statistics. Examples of such water masses in the surface layer of the North Pacific will be discussed in Section 3.4.

3.3.2 Recent view of surface water masses

Aspects of large-scale general circulation in the ocean differ greatly between the surface layer above several hundred meters and the deep layer

below. In the surface layer, wind-driven circulation is dominar
velocity is relatively large. On the other hand, large-scale curr
layer where thermohaline circulation dominates are
Consequently, the traditional method of calculating a geostr
assuming an appropriate level of no motion (normally a ρⅼₑₛₛᵤᵣₑ
selected) may be often valid for describing a surface current. But the method
does not necessarily work for the weak current in the deep layer, because how
to select a level of no motion decisively affects the resulting pattern of the
geostrophic current. That is why deep-water circulation has been described based
on the distribution of water masses and the spatial variation of their
characteristics. Even recently when an increasingly large number of directly
measured current data either from moored and drifted buoys are available,
analysis of water masses still remains an important technique to study the deep-
layer circulation.

In the meanwhile, large-scale currents in the surface layer of the ocean have
been described on the basis of various data such as direct current measurements,
deflection of ships, etc., not to speak of information from geostrophic currents.
The characteristic feature of the surface circulation thus revealed has been
explained relatively well by the theory of wind-driven circulation. The water
mass analysis as a fundamental technique for estimating a current field has not
been so emphasized for the surface layer as for the deep layer. Nonetheless, even
in the surface layer, distribution of characteristic water masses and their spatial
variation provide us with useful information on the three-dimensional structure
of the large-scale circulation.

Sverdrup *et al.* (1942) classified, for the first time, the global seawater into
several water masses based on the shapes of *T–S* curves (for details, see
Masuzawa and Hasunuma, 1977). For instance, the surface waters in the North
Pacific are divided into West North Pacific Central Water (Fig. 3-3) and East
North Pacific Central Water, either of which is represented by a straight line on
a *T–S* diagram. This classification was only geographical in two dimensions,
without distinguishing water masses in the vertical direction.

Montgomery (1958), among others, examined the distribution of water
volumes for each temperature–salinity class and, paying attention to the mode,
defined water masses. Masuzawa (1969) applied a similar method to the surface
waters in the subtropical North Pacific and named the water mass corresponding
to its mode as Subtropical Mode Water. It is a part of the seawater constituting the
West North Pacific Central Water but an interesting point is that it was described
as a water mass distributed three-dimensionally in the surface layer. The
Subtropical Mode Water is formed in the surface mixed layer by convection in
winter, and is distributed far beyond the formation area. This topic will be
discussed in detail in Section 3.4. Observing the formation, movement and
dissipation of this water mass is nothing but observing a part of the process that
maintains the three-dimensional structure of the surface water. This process is
directly associated with the transport of heat and materials inside the ocean, not
to speak of the three-dimensional circulation in the surface layer. Other surface

ater masses, which contribute to the knowledge about the three-dimensional distribution of seawater, will also be discussed in Section 3.4.

Clarifying the processes of formation, movement and mixing of main water masses is attributed to considering how the present three-dimensional structure of seawater is maintained, in balance with air–sea interaction at the surface and ocean currents. Furthermore, the object of the research on water masses goes beyond the clarification of their stationary aspects from formation to dissipation. In recent years, temporal variations of formation and circulation processes of water masses have attracted attention as they are regarded as a manifestation of variations of an ocean–atmosphere-coupled system. Moreover, the possibility of an active contribution by variations of water masses to the variation mechanism of the ocean–atmosphere-coupled system is being discussed. The time scale from formation to dissipation of the surface water mass is considered to be from a few years to a few tens of years, which is much shorter than that for the deep water layer. The surface water formation reflects, more or less, the atmospheric conditions at that time. That is, a certain change in the atmosphere might be transferred to a rather large region in the ocean in a few years or a few tens of years through the formation and circulation of a water mass. In addition, the resulting change in the ocean might act on the atmosphere in turn. The research on water masses from such a viewpoint is still in the beginning stage, but will become more and more important in the future.

3.3.3 Identifying and tracking of water masses

Masuzawa and Hasunuma (1977) systematically describe fundamental methods for identifying and tracking water masses. In this section, recent techniques will be discussed.

a) Temperature, salinity, potential temperature and potential density

The properties most used to identify water masses are temperature and salinity. This is not only because they are relatively easy to measure, but also because they are conservative properties that do not vary inside the ocean unless mixing takes place. Generally, water mass formation is associated with exchange of heat and fresh water at the sea surface with the atmosphere. Neither the temperature nor the salinity acquired in this process will be changed after the water is carried away from the formation area at the surface, as long as no mixing takes place.

Strictly speaking, temperature is not a conservative property, because it varies depending on pressure. Hence, potential temperature, which can be treated as a strict conservative property, is widely used. When we simply say potential temperature, it normally means the temperature of seawater raised to the sea surface adiabatically, and is expressed as θ. In general, θ can be obtained by integrating the adiabatic lapse rate, which is a function of water temperature, salinity and pressure, from the actual pressure to the sea surface pressure. A polynomial calculating the value of θ from temperature, salinity and pressure was devised (Bryden, 1973) and has been broadly in use. In contrast to θ, the temperature as measured at a given depth under the influence of pressure is called *in situ* temperature. The difference between θ and the *in situ* temperature is small (usually less than 0.1°C) for surface water masses from a depth of several hundred meters, so that the two can be regarded as equal for practical purposes. However, the strictly conservative property θ is more often used as it can be calculated in a relatively simple way, especially when temperature and salinity data are readily available.

In a broad sense, a potential temperature is the temperature of a water sample brought to any standard pressure surface. For example, potential temperatures based on the standard surfaces at 1000 db and 3000 db are often written as θ_1 and θ_3, respectively. The algorithm for calculating potential temperatures based on an arbitrary pressure surface is established. The method is summarized, together with the algorithms for obtaining different characteristics of seawater, in a technical report by UNESCO (1983).

The unit for temperature and potential temperature is "°C." For discussing temperature difference, we often use "deg" or "K." Upon a recommendation from the International Weights and Measures Committee, we have started to use the new water temperature scale ITS-90 (International Temperature Scale of 1990): the boiling point of water, which used to be 100°C, is now redefined as 99.974°C. A temperature value on the old temperature scale IPTS-68 (International Practical Temperature Scale of 1968) can be multiplied by 0.99976 to be approximately equal to a new value on the ITS-90. The difference is too slight to affect the analysis of surface water masses. Since no algorithm to calculate salinity or density based on the ITS-90 has been proposed so far, the IPTS-68 is still used for calculating salinity and density (Kawabe and Kawasaki, 1993).

Salinity was originally defined as concentration of all dissolved compounds in seawater with the unit ‰ (per mille). However, now salinity is generally obtained by measuring electric conductivity of seawater and the practical salinity scale, which is defined according to electric conductivity, has been officially adopted (Lewis and Perkin, 1978; Lewis and Fofonoff, 1979; UNESCO, 1981a). Salinity by this definition is called practical salinity and is normally the meaning of the term. It has no unit, but in most cases it is expressed with psu (practical salinity units) at the end.

According to the result of a detailed study on the difference between conventional salinity and practical salinity (Lewis and Perkin, 1981), they differ by less than 0.01 psu in normal seawater, which is negligible as long as the analysis of surface water masses is concerned. However, the difference is larger than the accuracy (± 0.003 psu) for present salinity measurements, so that it may be necessary to convert conventional salinity to practical salinity for certain cases, especially when deep water is involved in discussion.

The density (ρ) of seawater is a function of the *in situ* temperature, salinity and pressure: it can be calculated in the form of a polynomial (Millero *et al.*, 1980; UNESCO, 1981b). The density in the open ocean lies within a range of 1020–1070 kg·m^{-3} (sea surface–10000 db). For this reason, we use σ, instead of ρ itself, defined as

$$\sigma = \rho - 1000 \quad \left(\text{kg} \cdot \text{m}^{-3}\right). \tag{3-11}$$

The unit of σ is the same as that of ρ, but it is often written, for instance, as 27.5σ for the density 1027.5 kg·m^{-3}.

Seawater density is not a conservative property, because it varies depending on pressure. The density which seawater has under the pressure at a given depth is called *in situ* density. Potential density, which is a conservative property, can be obtained using a polynomial with replacing *in situ* temperature with potential temperature and replacing pressure with reference pressure. Potential density in a broad sense represents the density of seawater after it is brought adiabatically to any reference pressure surface. In a narrow sense, it is the density at the atmospheric pressure (namely at the sea surface), which is written as σ_θ. This potential density σ_θ is usually used for the analysis of surface water masses. As will be discussed later, for the analysis of deep water masses, a reference pressure surface of, say, 1000 db or 3000 db is selected as a representative depth of water, and the density of the water adiabatically brought to the pressure surface is used as σ_1 or σ_3. The potential density 1026.8 kg·m^{-3} referenced to the sea surface is written as $26.8\sigma_\theta$, and the potential density 1047.85 kg·m^{-3} referenced to the 3000 db surface is written as $47.85\sigma_3$.

Until recently a property called σ_t has been used extensively for water mass analysis. This is the density calculated from *in situ* temperature and salinity, but at zero pressure (atmospheric pressure) minus 1000 kg·m^{-3}. It would correspond to σ_θ, if θ were replaced by *in situ* temperature in the calculating formula (that is, it can be called the density of the water after it is isothermally brought to the sea surface). In the surface layer, say, above 1000 m, the difference between θ and *in situ* temperature is small, so that the difference between σ_θ and σ_t is also small (normally less than 0.02). Hence, σ_t can be read as σ_θ in practice.

Specific volume α ($=\rho^{-1}$), an inverse of density, is a widely used property because its vertical integration leads to a dynamic height (or geopotential thickness) in the ocean. For water mass analysis, it is not as often used as it used to be. The unit of specific volume is m^3·kg^{-1}.

In situ specific volume $\alpha_{s,t,p}$ is a function of salinity, *in situ* temperature and pressure. For the same reason as we use σ instead of ρ, we often use specific volume anomaly δ defined as

$$\delta = \alpha_{s,t,p} - \alpha_{35,0,p} . \tag{3-12}$$

The reference value $\alpha_{35,0,p}$ is the specific volume of seawater with salinity 35 psu, water temperature 0°C and pressure p: it is selected so that δ will have a positive value in a normal environment. The part of δ that can be determined only by temperature and salinity, and independent of pressure, has a special name "thermosteric anomaly" (Montgomery and Wooster, 1954) and it is written as δ_T or $\Delta_{s,t}$. It is the specific volume anomaly obtained from *in situ* temperature and salinity and zero pressure. δ_T corresponds to σ_t: one can be converted to the other (Pickard and Emery, 1990).

b) T–S diagram

As discussed in Section 3.3.1, the T–S diagram is a fundamental tool for identifying a water mass by a combination of temperature and salinity. If θ is used instead of temperature, the correct name should be a θ–S diagram, but we often call this a T–S diagram also (Fig. 3-3). σ_θ at an arbitrary point on a θ–S diagram is uniquely determined. Then it is convenient to plot isoplethic curves of σ_θ on the θ–S diagram as well (for the T–S diagram, isopleths of σ_t) as shown in Fig. 3-3.

The θ–S diagram (or T–S diagram) is useful not only for classifying water masses but also for studying the mixing of water. How water mixing appears on a θ–S diagram will be accordingly explained using a simple example. Suppose that three homogeneous water masses overly one another. We will name the three layers as A, B and C from top to bottom. We will assume that B and C have the same temperature but different salinities. Temperature and salinity profiles are shown in Figs. 3-4(a) and (b), while the θ–S relationships are shown in part (c). The diagrams at the top illustrate the situation before any mixing has occurred, while those in the middle and at the bottom show subsequent stages as mixing progresses. Initially the three water masses are homogeneous and may be represented on the θ–S diagram by three points (i.e., they are water types). As mixing progresses the sharp interfaces between the water masses become transition zones and the characteristics of water changes continuously. Since temperature and salinity are both conserved properties, the characteristics of the water formed by mixing of A and B are represented by the straight line connecting A and B. Likewise, the water formed by mixing of B and C is on the straight line between B and C. Further mixing leads to losing the identification of the original B layer, creating a water mass which is a mixture of A, B and C. The characteristics of the new water lie within the triangle ABC on the θ–S diagram (bottom). From this, we can tell that a θ–S curve like the one in this θ–S diagram generally takes place when three kinds of water masses are mixed. Not many cases are as simple as such a hypothetical scenario, but θ–S diagrams provide useful suggestions for more complicated mixing processes of water masses.

c) Potential vorticity

Potential vorticity is a dynamical conservative property used as a tracer of water masses. It is conserved unless dynamical or thermo-dynamical forcing or mixing exists. For a homogeneous ocean of uniform density, potential vorticity q can be written as

$$q = \frac{f + \zeta}{h} \qquad (3\text{-}13)$$

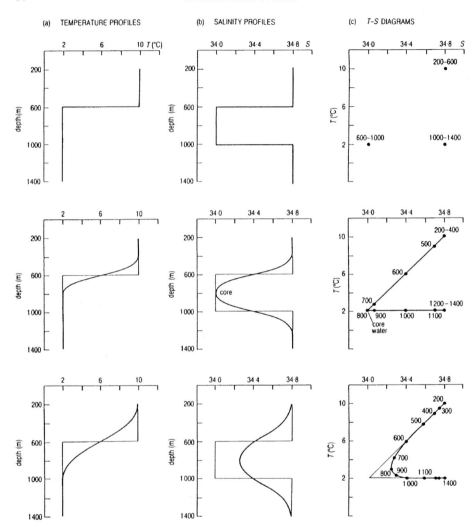

Fig. 3-4. Profiles of (a) temperature and (b) salinity, along with (c) the corresponding
T–S diagrams to illustrate the mixing of three homogeneous water masses (water types).
Stage 1 (top panel) represents the situation before any mixing has taken place; stage 2
(middle panel) shows an early stage of mixing when the core of intermediate water
is very prominent; by stage 3 (bottom panel), the core has been eroded (Open
University, 1989).

where f is the Coriolis parameter, ζ is the vertical component of relative vorticity,
and h is the thickness of the fluid layer. $(f + \zeta)$ represents the vertical component
of absolute vorticity. The conservation of potential vorticity $(Dq/Dt = 0)$ may be
understood as a consequence of the conservation of circulation $(D(f + \zeta)ds/Dt = 0)$
of a water column with an infinitesimal area ds and the conservation of its volume

(*Dhds/Dt* = 0).

In the region where relative vorticity is small compared with planetary vorticity ($f \gg \zeta$), approximate estimation of potential vorticity can be made using information of the density field, namely, temperature and salinity data. Potential vorticity q for the surface ocean made of continuously stratified fluid is often calculated by the following formula, when relative vorticity is negligible.

$$q = \frac{f}{\rho} \cdot \frac{\partial \sigma_\theta}{\partial z} \qquad (3\text{-}14)$$

where σ_θ is the potential density referenced to the sea surface. In general, vertical gradients of σ_1, σ_3, etc. are used to calculate corresponding potential vorticities. In addition, a method using the gradient of potential density referenced to local pressure was devised (Keffer, 1985).

We can use potential vorticity as a good tracer when we try to track a water mass that is formed by deep vertical convection. Such a water mass acquires low potential vorticity, which is conserved after the water mass is carried away from the sea surface, and consequently we can identify this water mass as a vertical minimum of potential vorticity. An example is illustrated in Fig. 3-5. In the figure, the vertical section of potential temperatures observed along the 137°E line by the Japan Meteorological Agency in the summer of 1986 are plotted with

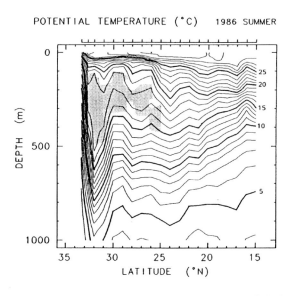

Fig. 3-5. Potential temperature section along 137°E in the summer of 1986. The stippled area represents a layer whose potential vorticity is lower than 2.0×10^{-10} m^{-1}·s^{-1} (Suga *et al.*, 1989).

potential vorticity less than 2.0×10^{-10} $m^{-1} \cdot s^{-1}$ indicated by shaded areas. The Subtropical Mode Water discussed in Section 3.2.2 corresponds to the water mass at the vertical minimum of this potential vorticity.

The use of the potential vorticity, rather than the vertical density gradient, as a tracer for water masses formed by deep vertical convection can be justified based on dynamical consideration, and so it should be used as long as temperature and salinity data are available. On the other hand, in the region such as subtropical surface layers where salinity contributes less to density variations than temperature does, the vertical temperature gradient, instead of potential vorticity, is sometimes used, especially when only temperature data from XBTs are available. This method is approximately valid except when we need to track a water mass for a wide range of latitudes.

d) Oxygen, AOU and nutrients

Non-conservative properties in the ocean, which vary through bio-chemical processes, provide useful information on the origin, movement, and mixing processes of water masses, if their source and sink characteristics are carefully handled. Dissolved oxygen and nutrients are good examples.

Among various dissolved substances, oxygen is most widely used for water mass analysis. It is measured most often next to temperature and salinity. The content of dissolved oxygen is almost saturated when water is in contact with air. The amount of saturation depends on temperature and salinity in the water. The relationship between the saturated dissolved oxygen and temperature/salinity is given by a polynomial (Weiss, 1970). Oxygen is consumed by organic decomposition and other bio-chemical processes, and eventually decreases after water loses contact with air. Thus it can be used as an index of the elapsed time since the water mass is formed near the sea surface. A more direct index in wide use is AOU (apparent oxygen utilization), which is the difference between the saturated amount of oxygen and the amount of dissolved oxygen. AOU increases as the time increases after water leaves the sea surface, but the rate of the increase depends on the region and the depth, and also the effect of mixing, so that AOU cannot be converted easily to the absolute time. Nonetheless, it is valid to examine the relative "age" of water masses that went through similar processes of formation and movement. Concrete examples will be given in Section 3.4.

Nitrogen, phosphorus and silicon are the three elements that are essential for plants to make organic matter from inorganic matter, and that tend to be lacking in seawater. They are limiting factors or "fertilizers" for plant production in the sea and consequently are called nutrients. The nutrients are used for water mass analysis, playing an important role, especially for identifying and tracking water masses in mid and deep layers. Normally, nitrate, phosphate and silicate are included in measurements in the open sea. Although measuring these will take more time compared with temperature, salinity and oxygen, they provide us with useful information on movement and mixing of water masses, if knowledge of their source and sink are available, in the same way as oxygen does. More detailed discussion on nutrients is given by Tsunogai and Noriki (1983).

e) Isopycnal surface analysis

It is most likely for seawater to flow and mix along a neutral surface. That is, the dominant direction is selected so that water particles are not subject to buoyancy force when they move along its surface. Accordingly, it is quite effective to examine the distribution of water properties and the flow on the neutral plane for studying the large-scale circulation and mixing of water masses. In practice, we regard potential density surfaces of an equal value as neutral surfaces and study the distribution of water properties, etc. on these surfaces. This is called isopycnal surface analysis. A typical example of applying isopycnal surface analysis to actual data is drawing a "horizontal" distribution chart of water properties on an isopycnal surface. Since advection and mixing prevail on this surface, it is easier to interpret than a horizontal distribution chart of water properties on an isobathymetric surface or an isobaric surface. As an axis of ordinates for the vertical section, potential density instead of depth or pressure is sometimes used. In recent years, observations of CTD sections in high resolution (short observation distance) have been made possible. The distinct structure of water masses is being clarified by plotting water properties on isopycnal surfaces based on such observations.

It may be worthwhile to note that water properties on isopycnal surfaces are studied together with the distribution of geostrophic streamlines on them. As will be described later, geostrophic streamlines on isopycnal surfaces are approximated by acceleration potential.

The potential density σ_θ referenced to the sea surface can be used for the isopycnal surface analysis of water masses in surface layers, but in general a particular reference surface for potential density should be carefully selected. In principle, it is best to use the reference pressure that is near the depth of the water mass to be analyzed (Lynn and Reid, 1968; Reid and Lynn, 1971). For instance, σ_θ is most appropriate for water masses between 0 and 500 m, and σ_3 (potential density referenced to 3000 db) for those between 2000 and 4000 m. This is because the thermal expansion coefficient $\alpha^{1)}$ and the contraction coefficient for salinity β depend on pressure: the following illustration about the strict definition of a neutral surface will help understand the reason.

A neutral surface in a strict sense can be defined as the surface n that satisfies the relationship (McDougall, 1987)

$$\alpha(p)\nabla_n\theta_p = \beta(p)\nabla_n S \qquad (3\text{-}15)$$

where θ_p is potential temperature referenced to the pressure p, S is salinity and ∇_n is a "horizontal" derivative along the neutral surface. This relationship shows that the density change due to the temperature change along the neutral surface compensates the density change due to salinity.

---

[1]Note that α here is not the specific heat, unlike in other sections of this chapter.

The potential density surface referenced to the pressure p_r is defined as the surface σ_r that satisfies

$$\alpha(p_r)\nabla_\sigma\theta_r = \beta(p_r)\nabla_\sigma S \qquad (3\text{-}16)$$

where θ_r is the potential temperature referenced to the pressure p_r, and ∇_σ is the "horizontal" derivative along the equi-potential density surface, that is, isopycnal surface σ_r. Comparing Eq. (3-15) and Eq. (3-16), we can tell that the potential density surface satisfactorily approximates the neutral surface as long as the *in situ* pressure is close to the reference pressure p_r. In other words, the nearer the reference surface is to the water mass in consideration, the better approximated the neutral surface is by the isopycnal surface (see Fig. 3-6).

f) Analysis using extremal values

Another method that is used as often as the isopycnal surface analysis is the analysis with special attention to extremal values in the vertical profile of water properties. For instance, a minimum of salinity is selected instead of an isopycnal surface and its depth, the distribution of water properties on the minimal surface, etc. are studied. This method is also called the core analysis (Wüst, 1935),

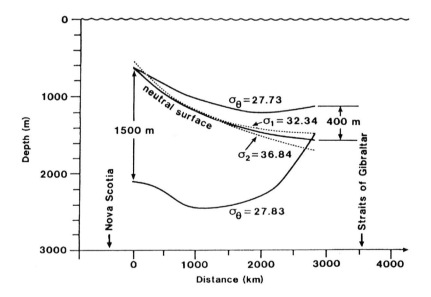

Fig. 3-6. Cross section, which goes from near Nova Scotia on the left to near the Straits of Gibraltar on the right, illustrates that the potential density surfaces of $27.73\sigma_\theta$ and $27.83\sigma_\theta$ intersect the same neutral surface at different positions. Also shown (dashed lines) are a potential density surface referenced to a pressure of 1000 db ($32.34\sigma_1$) and a potential density surface referenced to 2000 db ($36.84\sigma_2$) (McDougall, 1987).

because when the movement and mixing of a water mass are to be identified by a vertical minimum or maximum of water properties, the surface of the minimum in the properties is supposedly least subject to mixing, namely, it corresponds to the core of the mass.

For water masses that are illustrated by vertical extremal values of water properties, their spatial structures, which were not known by existing bottle sampling data, are now clarified using high-resolution CTD data in the vertical direction. Those water masses that were unidentifiable are also distinguished. Such examples will be described in Section 3.4.

Generally speaking, by analyzing the surface of vertical extreme values that represents the core of a certain water mass together with the analysis of the isopycnal surface that corresponds to the surface of governing advection and mixing, we have a better understanding of their movement and mixing.

g) Acceleration potential

Recent years have seen much progress in measuring ocean currents directly. Nevertheless, estimation of the current field from temperature and salinity data assuming geostrophic balance is still an important method for the study of three-dimensional flow of the ocean.

Geostrophic balance can be expressed as equilibrium between the Coriolis force and the pressure gradient force on a horizontal plane (equi-geopotential surface):

$$k \times f v = -\alpha \nabla_\phi p \qquad (3\text{-}17)$$

where k is the vertical unit vector, v the geostrophic velocity vector, α the specific volume, p the pressure, ϕ the geopotential, and ∇_ϕ the derivative along the equi-geopotential surface.

Let's consider a surface s of equal values of any scalar, and its derivative ∇_s along the surface, then

$$\nabla_\phi p = \nabla_s p - \frac{\partial p}{\partial \phi} \bullet \nabla_s \phi. \qquad (3\text{-}18)$$

With this and the equation of hydrostatic balance $(\partial p / \partial \phi = -1/\alpha)$, we can write the geostrophic relationship as

$$k \times f v = -(\alpha \nabla_s p + \nabla_s \phi) \qquad (3\text{-}19)$$

$$= -(\delta \nabla_s p + \nabla_s \phi_a) \qquad (3\text{-}20)$$

where δ is specific volume anomaly. ϕ_a is geopotential anomaly and can be written as

$$\phi_a = \phi_0 - \int_{p_0}^{p} \delta \, dp \qquad (3\text{-}21)$$

where ϕ_0 is the geopotential anomaly on the reference pressure surface p_0.

If the surface δ is selected as the surface s, then the geostrophic relation will be

$$k \times fv = -\nabla_\delta (\phi_a + \delta p) \qquad (3\text{-}22)$$

$(\phi_a + \delta p)$ is called acceleration potential, and its derivative on the δ surface represents geostrophic current (Montgomery, 1937; Montgomery and Stroup, 1962).

In practice, it is often assumed that $\phi_0 = 0$ (p_0 being chosen as the level of no motion), so that either of the following equations is used in most cases.

$$\phi_a + \delta p = -\int_{p_0}^{p} \delta \, dp + \delta p \qquad (3\text{-}23)$$

$$= \int_{\delta_0}^{\delta} p d\delta + \delta_0 p_0 \, . \qquad (3\text{-}24)$$

When we consider geostrophic currents on an isopycnal surface, we can assume that δ varies very little on the surface. Then we can generally use $(\phi_a + \delta p)$ for the acceleration potential on the potential density surface. However, more strictly, Eq. (3-20) will be

$$k \times fv = -\left[\nabla_\sigma (\phi_a + \delta p) - p \nabla_\sigma \delta \right] \qquad (3\text{-}25)$$

where ∇_σ is the derivative along the equi-potential density surface. Hence, the neglected term $-p \nabla_\sigma \delta$ on the right-hand side is the error. Comparing this error term with $\nabla_\sigma (\phi_a + \delta p)$ makes it possible to check the validity of the approximation (Zhang and Hogg, 1992).

h) *Making averaged fields*

A synoptic data is one made of nearly simultaneous observations in a broad area, such as the data from a series of stations along an observation line. When we conduct water mass analysis based on observations, we normally analyze synoptic data. But synoptic data contain phenomena whose spatial and temporal scales are smaller than those in consideration, so that it is sometimes difficult to interpret them. Also synoptic data at a particular time might not be enough for covering the whole broad area we want to analyze.

When we study characteristics of the large-scale distribution, formation, and movement of a water mass, in many cases it is effective to make averaged fields. There is a variety of methods for making averaged fields: spatial smoothing of synoptic data, time-averaging of data from a fixed place or from a certain area, averaging with respect to both time and space, etc. It is necessary to select the best method depending on the purpose of the analysis and the availability of data. In general there are two things to be noted as described below.

It is sometimes effective to take the potential density for the vertical axis for time-averaging or smoothing in space. This is because the advection and mixing of water masses are dominant along the isopycnal surface. When the depths of the density surface become wavy due to a dynamical disturbance on the meso-scale or smaller, we may erroneously average properties on different density surfaces if we take the depth as the vertical axis.

When we make the average field of a structure with large temporal variations and with a front in water properties such as the Kuroshio, the distance from the front is sometimes chosen as the coordinate. This method is appropriate for examining the average structure near the front. Analyses based on such a method were made for the Gulf Stream area (Halkin and Rossby, 1985) and for the Kuroshio and its extension areas (Hanawa and Hoshino, 1988; Bingham, 1992). Figure 3-7 shows an example of a long-term averaged cross section of the Kuroshio area. The upper figure is made of a simple average using geographical coordinates, whereas the lower figure is the result of the average taking the distance from the Kuroshio axis as the "north-south" coordinate. The latter captures the temperature structure near the Kuroshio more sharply.

3.4 DYNAMICAL INTERPRETATION OF FORMATION OF SURFACE WATER MASSES AND SOME EXAMPLES

In this section the results of recent studies on typical water masses in the surface layers of the subtropical North Pacific will be described.

Attempts to interpret dynamically the formation of water masses in the surface layers have made much progress in recent years. The basic concept of the dynamical interpretation of water mass formation will be discussed first.

3.4.1 Formation of surface water masses and ventilation

Ventilation is a process in which seawater with given initial characteristics of the mixed layer is sent to the lower layer, namely, it is the process of seawater in contact with the atmosphere entering into the lower layer. Ventilation is associated with not only the formation of surface water, but also with the formation of all water masses in mid, deep and bottom layers. The ventilated thermocline theory deals with the ventilation of the main thermocline due to Ekman pumping in the subtropical gyre (Luyten et al., 1983). This ventilation

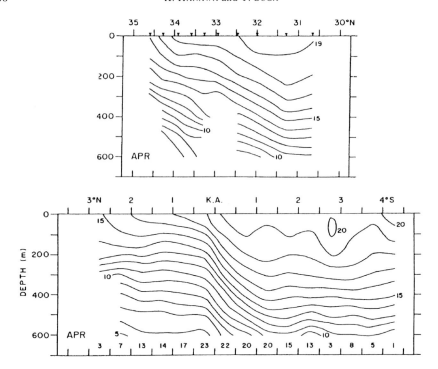

Fig. 3-7. Long-term mean temperature vertical section (unit in °C) over the Izu Ridge for April. The upper panel is the mean field obtained by simple averages for individual depths of each section. Small triangles show the observational stations. The lower panel is the mean field obtained by averages taking the distance from the Kuroshio axis (K.A.) as the "north-south" coordinate. Numerals in the bottom of the figure show the number of the data for each box (Hanawa and Hoshino, 1988).

associated with the pushing down of seawater into the thermocline by Ekman pumping is called subduction. The subduction rate, or the speed of the downflow of seawater in the mixed layer into the lower layer, was estimated from the climatological density field of the ocean and the climatological sea surface fluxes (e.g., Marshall $et\ al.$, 1993).

The concept of the subduction rate may be explained as follows (see Fig. 3-8). Let h be the thickness of the mixed layer, $\boldsymbol{u}_b$ and w_b be the horizontal and vertical current velocities at the lower boundary of the mixed layer, respectively. Then the subduction rate S at a certain instance will be

$$S = -\frac{\partial h}{\partial t} - \boldsymbol{u}_b \bullet \nabla h - w_b \qquad (3\text{-}26)$$

S represents the volume of the water entering from the mixed layer to the thermocline per unit area per unit time, or the volume flux of seawater passing

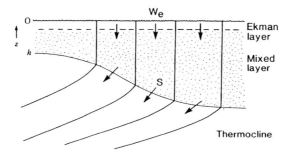

Fig. 3-8. A schematic diagram showing isopycnals in the thermocline outcropping into a vertically homogeneous mixed layer driven by buoyancy fluxes and Ekman pumping at its surface. Quasi-horizontal flow in the mixed layer slides into the thermocline through the sloping base of the mixed layer. The mass flux per unit surface area through the mixed-layer base is S; it is the field of S that ventilates the thermocline (Marshall et al., 1993).

through the lower boundary of the mixed layer.

An example of estimating the subduction rate using actual data is given next. The thickness and water properties of the mixed layer vary season by season and the corresponding different waters are pushed down into the lower layer. According to Stommel (1979), among these waters, the one that is pushed deeper than the thickest mixed layer in winter is solely responsible for the subduction of the main thermocline. Let H be the thickness of the mixed layer in winter, and assuming there is no interannual variations so that the time derivative is neglected, the annual subduction rate S_{ann} of the seawater being pushed downward through the surface $z = -H$ is given by

$$S_{ann} = -\overline{w_H} - \overline{u_H} \bullet \nabla H . \qquad (3\text{-}27)$$

The subscript H represents the value at $z = -H$, and the overbar denotes the annual average. It should be noted that w_H is not necessarily equal to the Ekman pumping velocity w_{Ek}. If there is north-south velocity v above $z = -H$, and the linear vorticity balance ($\beta v = f \partial w/\partial z$, here $\beta = \partial f/\partial y$) holds, then

$$w_H = w_{Ek} - \frac{\beta}{f} \int_{-H}^{0} v \, dz . \qquad (3\text{-}28)$$

In this case, the annually averaged subduction rate can be estimated by the vertical velocity at the lower boundary of the mixed layer (the first term on the right-hand side of Eq. (3-27), vertical pumping), and the horizontal velocity passing the lower boundary of the mixing layer (the second term on the right-hand side of Eq. (3-27), lateral induction).

According to a study on such an estimation of the subduction rate using the stress field of surface wind and climatological data of the density field in the ocean, the lateral induction is sometimes larger than the vertical pumping in certain areas. Although the annually averaged subduction rate is based on some simplifications and assumptions, it gives an important basis for the dynamical interpretation of the formation process of water masses and quantifying their formation rate.

3.4.2 North Pacific Subtropical Mode Water

In a broad area of the northwestern part of the North Pacific subtropical gyre, there is a water mass characterized by the minimal layer of the vertical temperature gradient, or thermostad, in the depth of 100–400 m. Since salinity in this layer is almost uniform, the water mass is also the minimal layer of the vertical density gradient (pycnostad, see Fig. 3-5). As described in Section 3.2, the water properties of this water mass constitutes the mode of the volume distributions of temperature and salinity classes in the subtropical surface layers; it is named Subtropical Mode Water (Masuzawa, 1969).

A comparable water mass exists in the North Atlantic subtropical gyre, and its typical temperature is 18°C; it was called Eighteen Degree Water (Worthington, 1959). There is also a similar water mass in the South Pacific (Roemmich and Cornuelle, 1992). Now these waters are called North Pacific Subtropical Mode Water, North Atlantic Subtropical Mode Water, and South Pacific Subtropical Mode Water, respectively. In this section only North Pacific Subtropical Mode Water will be discussed and it will be called simply Subtropical Mode Water.

The Subtropical Mode Water is formed by deep convection due to cooling in winter just to the south of the Kuroshio Extension area. The formation area is where the heat release is the largest in the world oceans, due to a huge amount of heat brought by the Kuroshio and the effect of cold outbreak from the Asian continent (East Asian monsoon). In other words, the strong interaction between the ocean and atmosphere in winter generates Subtropical Mode Water. This water mass has a significant vertical uniformity of water properties and is characterized by the vertical minimum of the potential vorticity.

When the sea surface is heated in spring and after, Subtropical Mode Water is cut off from the atmosphere and stays in the subsurface layer as the thermostad or the pycnostad. It is then carried away from its formation area by the anticyclonic flows in the Kuroshio recirculation spreading into the recirculation region. The annual mean acceleration potential with respect to 1000 db on the $25.4\sigma_\theta$ surface is shown in Fig. 3-9(a) for the non-large-meander period and in Fig. 3-9(b) for the large-meander period of the Kuroshio. This surface lies within the pycnostad of Subtropical Mode Water. During the non-large-meander period, an intense anticyclonic circulation appears south of Japan. The anticyclonic gyre includes the southwestward Kuroshio Countercurrent, the typical geostrophic current velocity of which is 5–10 cm/sec southwestward.

Figure 3-10 represents isopycnal maps, for each season during the non-

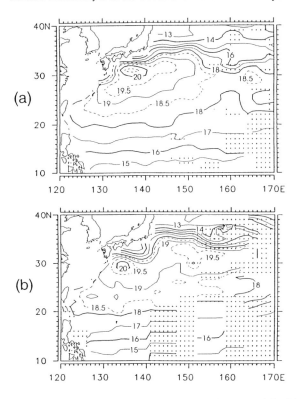

Fig. 3-9. Climatological maps of annual mean acceleration potential with respect to 1000 db on the $25.4\sigma_\theta$ surface in (a) the non-large-meander period and (b) the large-meander period of the Kuroshio. Units are $m^2 \cdot s^{-2}$. Crosses indicate grid points where the 95% confidence intervals of the mean are beyond a typical contour interval, $0.5 \ m^2 \cdot s^{-2}$ (Suga and Hanawa, 1995a).

large-meander period of the Kuroshio, of potential vorticity on the $25.2\sigma_\theta$ surface which corresponds to Subtropical Mode Water. It is seen that a significant lateral minimum of the potential vorticity exists in the Kuroshio Extension area in winter. This indicates the formation area of the water mass at this density. In spring and the following seasons, the minimum shifts southwestward as expected from the geostrophic flow pattern (Fig. 3-9(a)), while the value of the potential vorticity increases. That is, the newly formed water mass is advected by the southwestward flow and is gradually dissipated. This figure is based on climatological data and exactly the same variation pattern does not necessarily occur every year, but since the temporal scale from the formation through movement and dissipation is less than a year or so, those processes are captured as seasonal variations on the average.

    The seasonal variation of Subtropical Mode Water as a manifestation of formation, movement or dissipation can be observed by means of oxygen, which

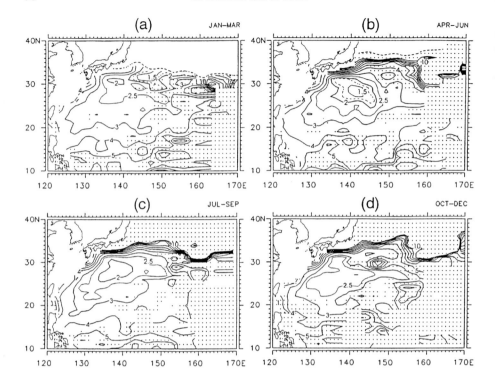

Fig. 3-10. Climatological maps of potential vorticity on the $25.2\sigma_\theta$ surface during the non-large-meander period of the Kuroshio for (a) winter, (b) spring, (c) summer, and (d) autumn. Units are 10^{-10} $m^{-1}\cdot s^{-1}$. Crosses indicate grid points where the 95% confidence intervals of the mean are beyond a typical contour interval, 0.5×10^{-10} $m^{-1}\cdot s^{-1}$ (Suga and Hanawa, 1995a).

is a non-conservative quantity. Figure 3-11 shows contour maps of sectional area distribution of long-term mean potential temperatures and AOU based on the observations along the 137°E-line, which are made routinely by the Japan Meteorological Agency. For summer (Fig. 3-11(a)), at potential temperatures higher than 17°C, a predominant ridge runs almost along the AOU value of 0.6 ml/l. In the winter subsurface diagram (Fig. 3-11(b)), a single ridge lies along the AOU value of 0.9 ml/l. The mixed layer water in winter (Fig. 3-11(c)) has potential temperature values higher than 18.6°C and AOU values around 0.3 ml/l, which are lower than those of the summer ridge. The areal-distribution mode around 0.6 ml/l in summer corresponds to Subtropical Mode Water which was formed to the east of this section in the previous winter (four or five months before) and brought here. The temperature of the AOU mode in winter is lower than that of the AOU mode in summer; this is the colder Subtropical Mode Water that was formed farther to the east a year before and brought here.

    The AOU value increases systematically from the winter mixed layer to summer and from summer to winter subsurface at an approximate rate of 0.6 ml/l/year. This increase does not represent the Lagrangian change of tracing the same water mass. However, the AOU has an almost constant value of 0–0.2 ml/l for all the formation area of Subtropical Mode Water (Suga and Hanawa, 1990), and the change in the figure may well represent how the AOU value increases after the water mass is formed.

There is no estimation of the AOUR (apparent oxygen utilization rate) for the subsurface layer in the North Pacific subtropical gyre, but for the subtropical North Atlantic, Jenkins (1980) estimated it to be 0.5 ml/l/year. This value corresponds well to the rate of increase shown in Fig. 3-11, and this indicates that the estimate mentioned above, 0.6 ml/l/year, is not unrealistic. However, the increasing rate 0.6 ml/l/year of the AOU in Subtropical Mode Water may well include not only the oxygen consumption by biochemical processes but also the effect of mixing. At any rate, the difference between the AOU value of the mode water formed this year and that formed last year is as large as 0.5 ml/l, so that it is relatively easy to assess the formation year of a subtropical mode water.

Formation and advection processes of Subtropical Mode Water vary interannually. As discussed before, its formation process is closely associated with the ocean–atmosphere interaction in winter. Oceanic cooling due to atmosphere changes year by year. The intensity of the subtropical gyre, or the heat transported by the Kuroshio, also varies. These should result in changes of the mass and characteristics of Subtropical Mode Water.

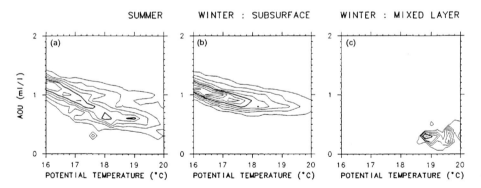

Fig. 3-11. Distribution of the area along 137°E in each bivariate potential temperature–AOU class $0.2°C \times 0.1$ ml·l^{-1} for all sections which were taken in summer from 1972 to 1986 and in winter from 1973 to 1987, within the potential temperature range of 16 to 20°C and the latitude range south of the Kuroshio axis to 20°N. No diagram is given for the summer mixed layer because its temperature is above 20°C. Contour lines are drawn for each 10 km^2 of area with thickening of each 50 km^2 (Suga et al., 1989). Note that AOU in this figure is biassed by approximately 0.18 ml·l^{-1} because it is calculated with Fox's (1909) polynomial of the saturated dissolved oxygen instead of Weiss' (1970).

A measure of the intensity of cooling due to monsoon often used is the wintertime mean monsoon index (MOI), which is defined as the sea level pressure difference between Nemuro, Japan and Irkutsk, Russia. This is because the wintertime cooling in the western North Pacific is thought to be dominantly controlled by the East Asian wintertime monsoon. The MOI represents the overall character of cooling in a wide area in a given winter. Figure 3-12 shows the relationship between the variation of the sea surface cooling and the MOI. The upper panel (a) represents the time series of the MOI, and the lower panel (b), the distribution of coefficients of correlation between MOI and sea surface temperature in winter. It is demonstrated that high MOIs are associated with low sea surface temperatures representing the sea surface cooling in the northwestern part of the subtropical gyre.

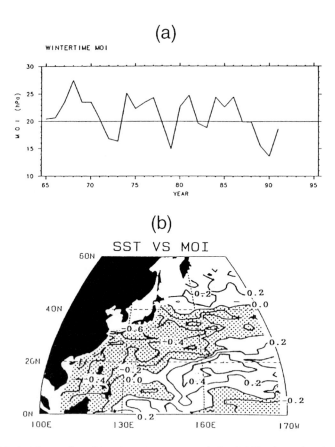

Fig. 3-12. (a) Time series of the wintertime monsoon index (MOI): the sea level pressure difference between Nemuro, Japan and Irkutsk, Russia, in hPa. (b) The distribution of coefficients of correlation between the MOI and sea surface temperature (SST) in winter. The stippled areas represent negative correlation coefficients (Suga and Hanawa, 1995b).

A latitude–time diagram of potential vorticity and AOU on the $25.3\sigma_\theta$ surface, lying within Subtropical Mode Water pycnostad, along the 137°E repeat section is presented in Fig. 3-13. Lower potential vorticity and lower AOU imply thicker and newer Subtropical Mode Water, respectively. Lower values of potential vorticity tend to be accompanied by lower values of AOU, which is confirmed by good positive correlation between potential vorticity and AOU at latitudes of typical Subtropical Mode Water distribution (not shown). This observation supports the view that low potential vorticity corresponds to water formed by vertical convection, which also lowers its AOU.

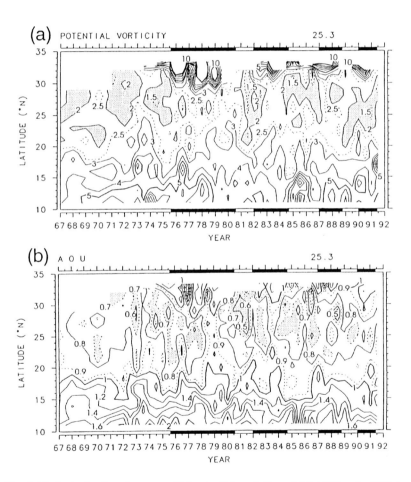

Fig. 3-13. Latitude–time diagrams of (a) potential vorticity and (b) AOU on the $25.3\sigma_\theta$ surface. Units are $10^{-10}\,\mathrm{m^{-1}s^{-1}}$ and $\mathrm{ml\cdot l^{-1}}$ for potential vorticity and AOU, respectively; potential vorticity lower than $2.0 \times 10^{-10}\,\mathrm{m^{-1}s^{-1}}$ and AOU lower than $0.7\,\mathrm{ml\cdot l^{-1}}$ are hatched. Thick lines top and bottom of figures indicate the large-meander period of the Kuroshio (Suga and Hanawa, 1995b).

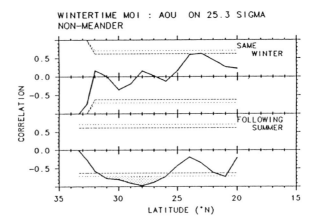

Fig. 3-14. Correlation coefficients at each station between the wintertime MOI and AOU on the 25.3σ_θ surface in the same winter and the following summer. Dotted lines and broken lines show the 95% and 90% confidence levels, respectively. Correlation coefficients beyond the 95% significant level are hatched (Suga and Hanawa, 1995b).

Figure 3-14 shows correlation coefficients between the wintertime MOI and AOU on the 25.3σ_θ surface in the same winter and the following summer, both during non-large-meander period. While no significant correlation is shown in the same winter, a notable negative correlation occurs at 26–30°N in the following summer. This suggests a strong relationship between the wintertime cooling and the renewal of Subtropical Mode Water. The correlation in the same winter is low because the just formed Subtropical Mode Water has not been advected to 137°E yet. Thus the advection of Subtropical Mode Water carries the results of wintertime ocean–atmosphere interaction to the wide area beyond the formation region.

If the memory of wintertime ocean–atmosphere interaction is carried by Subtropical Mode Water, then how does it affect the surface layer processes? This is a key question to understand the coupled ocean–atmosphere system which varies through the large-scale interactions of the two fluids. Unfortunately, we have not yet found a satisfactory answer.

It can be seen from Fig. 3-13 that there are fluctuations on a time scale longer than year-to-year changes. One such fluctuation is associated with the variation of the Kuroshio path. The Kuroshio has two typical paths south of Japan: one along the south coast of Honshu, the main island of Japan, the non-large-meander path, and the other detouring to the south off Honshu, the large-meander path (e.g., Kawabe, 1985). Much less Subtropical Mode Water at 137°E appears during the large-meander period. This is because the Kuroshio recirculation pattern changes during the large-meander period (Fig. 3-9(b)) so that the advection of Subtropical Mode Water from the Kuroshio Extension region to the south of Honshu is considerably reduced.

It should be mentioned that the change of the Kuroshio recirculation is not necessarily a cause and the change of Subtropical Mode Water is not necessarily a result. The latter may cause or encourage some changes in the recirculation. Subtropical Mode Water is considered as one of the low potential vorticity sources for the Kuroshio recirculation (Cessi, 1988) although its relative importance is not fully understood. The influence of Subtropical Mode Water fluctuations on recirculation will be an interesting future subject as it is a possible mechanism with which wintertime ocean–atmosphere interaction affects the surface circulation through a water mass process.

For studying longer-term changes of Subtropical Mode Water, Bingham *et al.* (1992) compared thermal conditions in two pentads, 1938–1942 and 1978–1982. Subtropical Mode Water was thicker, laterally more homogeneous in its temperature, and geographically more confined in its distribution during 1938–1942 than the other pentads (Figs. 3-15 and 3-16). The thicker mode water was attributed to more intense wintertime cooling during 1938–1942 based on comparison of the MOI between the two pentads. On the other hand, the overall upper thermal condition of the subtropical gyre suggests that the gyre was more spun up during 1978–1982 (not shown), which implies a stronger wind stress field over the subtropical North Pacific during this pentad. That is, the larger scale wind field rather than the East Asian monsoon may affect geographical distribution of Subtropical Mode Water: thus the wider distribution during 1978–1982. These discussions, though based on a limited comparison, may give a good starting point to clarify how the atmosphere, the ocean surface circulation and the

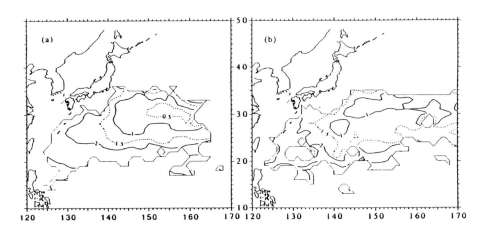

Fig. 3-15. Vertical temperature gradient on the surface of the vertical temperature gradient minimum corresponding to North Pacific Subtropical Mode Water for the pentads of (a) 1938–1942 and (b) 1978–1982. Light solid line encloses the area where the vertical minimum could be unambiguously identified. Units are °C/100 m (Bingham *et al.*, 1992).

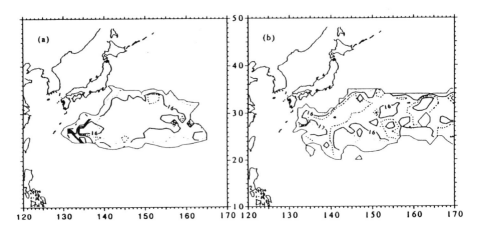

Fig. 3-16. Temperature on the surface of the vertical temperature gradient minimum corresponding to North Pacific Subtropical Mode Water for the pentads of (a) 1938–1942 and (b) 1978–1982. The heavy dashed contour is 16.5°C. For areas outside of the light solid line either the vertical minimum could not be unambiguously identified or the vertical temperature gradient is greater than 2.0°C/100 m. The light dashed line is the 1.5°C/100 m contour from Fig. 3-15 (Bingham *et al.*, 1992).

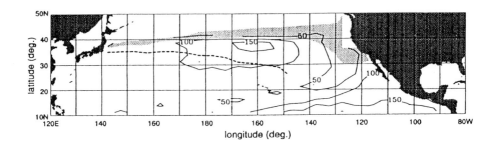

Fig. 3-17. Thickness (m) of the layer with temperature of 10–12°C, which approximately depicts thickness distribution of the North Pacific Central Mode Water core layer. Dashed line indicates the 12°C isotherm at 300-m depth representing the Kuroshio Extension path. Hatched zone indicates the zone where temperature at 300 m is between 6 to 8°C representing the Kuroshio Bifurcation Front in the western basin (Suga *et al.*, 1997).

Subtropical Mode Water formation/distribution interact with one another to cause variability of the coupled system.

3.4.3 North Pacific Central Mode Water

In addition to Subtropical Mode Water, there is another type of thermostad in the North Pacific. This is called North Pacific Central Mode Water (Suga *et al.*,

1997); thickness distribution of its core layer is approximately depicted as the thickness of the layer of 10–12°C (Fig. 3-17). The dashed line in the figure indicates the Kuroshio Extension path and the hatched zone the Kuroshio Bifurcation Front. Climatological maps (Fig. 3-18) of the wintertime mixed layer thickness together with the sea surface temperature distribution suggest that this water is formed in the deep mixed layer immediately south of the Kuroshio Bifurcation Front.

Talley (1988) examined potential vorticity distribution with the use of the Levitus (1982) climatological data. Her map of potential vorticity on the $26.2\sigma_\theta$ surface (figure 4 in Talley, 1988), a little denser than Subtropical Mode Water isopycnal, also shows a lateral minimum corresponding to North Pacific Central Mode Water. Nakamura (1996) further examined North Pacific Central Mode Water as a low potential vorticity water in the Levitus data and compared it to the lighter variety of the North Atlantic Subpolar Mode Water (McCartney and Talley, 1982).

The Central Mode Water, like Subtropical Mode Water, is formed as a result of the ocean–atmosphere interaction. Accordingly, its distribution and characteristics should vary as atmospheric forcing and oceanic circulation changes, but little work on its variability has been done. It is interesting to note

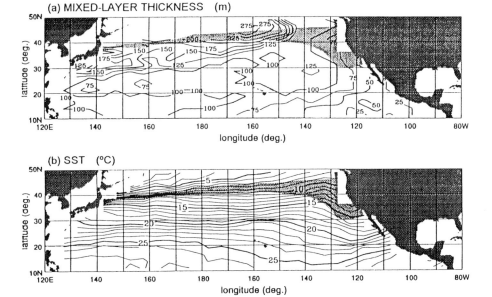

Fig. 3-18. Maps of wintertime (a) mixed layer thickness (m) and (b) sea surface temperature (°C). The hatched zone indicates the zone where temperature at 300 m is between 6 and 8°C representing the Kuroshio Bifurcation Front in the western basin (Suga *et al.*, 1997).

that its formation area nearly corresponds with the center of the decadal-scale sea surface temperature variability (Tanimoto *et al.*, 1993). It may be possible that the Central Mode Water conveys the decadal-scale fluctuations of the sea surface or the surface mixed layer to a broader area inside the ocean.

3.4.4 Tropical Water

High salinity water is formed near the Tropics due to excessive evaporation particularly in winter, so that salinity becomes maximal on the low latitude side of the subtropical circulation. Cannon (1966) named the corresponding water Tropical Water. It was called Tropical High Salinity Water by Masuzawa and Hasunuma (1977). Its characteristic values in the formation area are: 24°C, 35.5 psu, and $\delta_T = 390 \times 10^{-8}$ m^3/kg ($\sim24.0\sigma_t$) (Tsuchiya, 1970).

North Pacific Tropical Water is carried westward by the North Equatorial Current, and then part of it flows into the Kuroshio. Within the Kuroshio its salinity is lowered due to mixing (Masuzawa and Hasunuma, 1977), but it keeps its characteristics of the salinity maximum layer and can be traced up to the southern coast of Japan. Tropical Water is a source for high salinity in the subtropical gyre, but its formation and circulation processes have not been well described.

Temporal variation of Tropical Water has not been studied either, but at a particular meridian its variations on time scales from one year to several years were reported by Masuzawa and Nagasaka (1975). Figure 3-19 shows the year-to-year change of wintertime salinity distribution corresponding to Tropical Water from 1967 to 1974 at a cross section of 137°E. High salinity water north of about 10°N represents Tropical Water of the North Pacific. It seems that Tropical Water weakened after 1970 and recovered in 1974. Of course we cannot tell if this variation represents the whole of Tropical Water, but it suggests a possibility of large interannual variations in Tropical Water. It will be interesting to know how such variation is associated with variations of ocean–atmosphere interaction, circulation systems or characteristics of water masses in the surface layer.

3.4.5 Shallow salinity minimum

There are three layers of salinity minima in the North Pacific. One is defined to be North Pacific Intermediate Water, which will be discussed in Section 3.4.6. Another is known as Antarctic Intermediate Water. The remaining one is called a "shallow salinity minimum," which is observed in the eastern North Pacific south of 50°N and spreads over the eastern boundary current region and turns to the southwest around 25°N following the current (Fig. 3-20). At lower latitudes, it is found from 130°W to the western boundary between 10°N and the equator.

The origin of the shallow salinity minimum is surface water of low salinity and high oxygen concentration near 35–50°N, 145–160°W (Tsuchiya, 1982). This surface water is advected equatorward in the eastern part of the subtropical gyre and meets more saline but less dense subtropical surface water. The lower

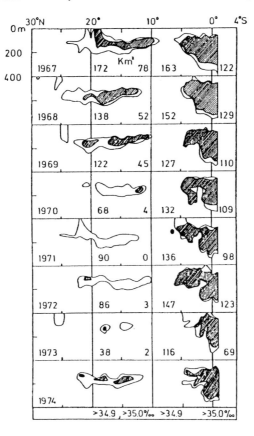

Fig. 3-19. Salinity maximum layer, more than 34.9 (contour) and 35.0 (hatched), at 137°E in January from 1967 to 1974. The high salinity layer north of about 10°N corresponds to North Pacific Tropical Water. The numerals in the figure denote the areas, km², more than 34.9 (left) and 35.0 (right) (Masuzawa and Nagasaka, 1975).

salinity surface water slides beneath the more saline surface water, which results in a subsurface salinity minimum. Talley (1985), using a modified version of ventilated thermocline model of Luyten *et al.* (1983), ascribed the formation of the shallow salinity minimum to a subduction process resulting from the wind-driven, ventilated circulation in the subtropical gyre. The results show that the shallow salinity minimum appears as a result of a certain combination of the density/salinity distribution in winter determined by thermohaline processes and the circulation pattern determined by wind forcing. The density and salinity of the minimum vary as the combination varies.

The distribution of salinity, using recent high-resolution CTD and STD data, is shown in Fig. 3-21. Note that three groups of minima along 152°W are plotted in the figure: North Pacific Intermediate Water (indicated by crosses), the

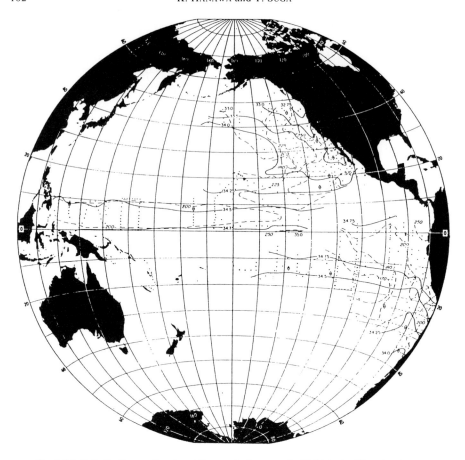

Fig. 3-20. Salinity (heavy lines) and thermosteric anomaly (light dashed lines) at the two shallow salinity minima derived from high latitudes. All the dots represent stations where the minima were observed. At many other positions within the area contoured (particularly within the eastern intertropical zone), other data did not clearly reveal a minimum, though they did not preclude it (Reid, 1973).

shallow salinity minimum (circles) and the "middle salinity minimum" (asterisks).

Figure 3-22 illustrates the relationship between density, salinity front and Ekman convergence. The $25.1–26.0\sigma_\theta$ isopycnals outcrop meridionally in a large northern region, north of the salinity front, of low surface salinity where there is Ekman downwelling. This outcropped line could be the source water area for the shallow salinity minimum (Yuan and Talley, 1992). On the other hand, a little denser surface of $26.0–26.5\sigma_\theta$ surface outcrops only in a rather restricted region of low salinity and downwelling. Accordingly, although this surface, which is located between the shallow salinity minimum and North Pacific Intermediate Water, produces the "middle salinity minimum", the minimum does

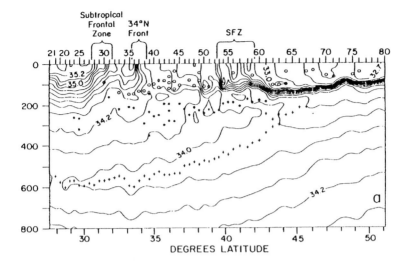

Fig. 3-21. Meridional section of salinity at 152°W. Crosses, asterisks, and circles indicate the North Pacific Intermediate Water, middle, and shallow salinity minima, respectively (Yuan and Talley, 1992).

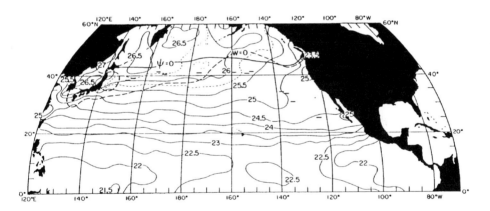

Fig. 3-22. Surface density in winter from Levitus' (1982) data. The heavy solid line represents the Sverdrup streamfunction $\psi = 0$ and the heavy dashed line represents Ekman pumping $W = 0$. Both are calculated from Hellerman and Rosenstein's (1983) winter (February, March, and April) climatological wind stress. Bars indicate the surface salinity front (Yuan and Talley, 1992).

not spread over a large area. While the shallow salinity minimum is created every winter, the "middle salinity minimum" is not formed in some years.

The shallow salinity minimum, together with North Pacific Intermediate Water, is a source for low salinity in the North Pacific. Its temporal variation of formation and distribution contributes to the variation of the salinity budget in the whole North Pacific. It will also be useful as a tracer for understanding water exchange between tropics and subtropics and its variability. We expect more studies on its temporal variations in the future.

3.4.6 North Pacific Intermediate Water

North Pacific Intermediate Water is recognized as a salinity minimum in the lowest layer of the wind driven subtropical gyre, which is not directly ventilated at the surface of the open ocean. In this section, this water mass is included in the surface layer in a broad sense.

Numerous researchers have studied North Pacific Intermediate Water as a well-defined salinity minimum at depths of 300–700 m in the subtropical North Pacific. Its formation and circulation have been interpreted in various ways. The history of the studies on North Pacific Intermediate Water from early years to the 1970s is well described by Masuzawa and Hasunuma (1977). Their description is summarized as follows.

Wüst (1929) and Uda (1935a, b) believed that low salinity water formed in the surface layer in the western subarctic in winter is gradually carried southward by intermediate currents or the Oyashio Undercurrent while sinking, and thus forming the salinity minimum. Sverdrup et al. (1942) had the same view except they considered the advection along the wind-driven circulation pattern from the horizontal distribution of the salinity minima.

Noting that salinity minimum lies almost along the isopycnal surface of $26.8\sigma_\theta$ ($\delta_T = 125 \times 10^{-8}$ $m^3{\cdot}kg^{-1}$), Reid (1965) regarded this potential density as a representative property of North Pacific Intermediate Water instead of salinity minimum itself. He showed that the circulation pattern of North Pacific Intermediate Water resembled the wind-driven surface circulation using the acceleration potential map on the isopycnal surface.

Reid further noted that the $26.8\sigma_\theta$ isopycnal surface does not outcrop in the North Pacific, that is, the subpolar surface water cannot be a direct source of North Pacific Intermediate Water. He hypothesized that a subsurface isopycnal surface of $26.8\sigma_\theta$ in the subpolar gyre acquires the characteristics of low temperature, low salinity and high oxygen through vertical mixing and then isopycnal mixing spreads these characteristics into the subtropical gyre to form the intermediate salinity minimum.

Hasunuma (1978) demonstrated that the salinity minimum is formed in the Oyashio–Kuroshio mixed water region, giving a thorough and important account of conditions there (Fig. 3-23). He indicated that the salinity minimum originates from an overrun of subpolar waters by subtropical waters. He then concluded that the intermediate salinity minimum, rather than being a convectively formed and hence volumetrically important water mass, is just the vertical boundary between subtropical and subpolar waters. Accordingly he proposed to refer to the salinity minimum itself as the North Pacific Intermediate Salinity Minimum, distinguished from the Intermediate Water.

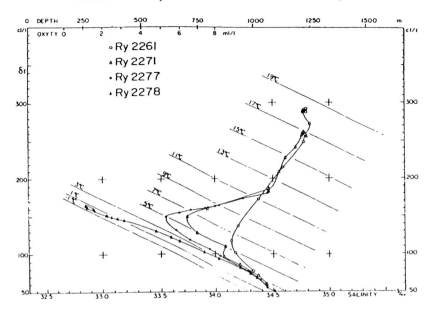

Fig. 3-23. δ_T-S curves which represent the vertical salinity structures at the northern and southern edges of the Kuroshio and Oyashio front at 144°E, February 1966 (Hasunuma, 1978).

Since the early 1990s new hypotheses about North Pacific Intermediate Water have been introduced. For example, Talley (1991) suggested that while vertical mixing does appear to freshen the $26.8\sigma_\theta$ surface in the subpolar gyre to some extent, most of the freshening and oxygenation take place directly through sea-ice formation in the Okhotsk Sea and through vertical mixing in the Kuril Straits. Talley (1993) regarded this process as "ventilation of an isopycnal," which should be distinguished from "formation of a salinity minimum at that density". She suggested that densest winter surface water sliding beneath or mixing laterally beneath the saltier surface layers is likely responsible for the latter process. She further showed that the only region where North Pacific Intermediate Water is formed in this sense is the northwestern subtropical gyre, that is the mixed water region between the Oyashio front and the Kuroshio Extension, by presenting spatial distribution of water properties at the salinity minimum (Fig. 3-24).

"Why does the density of the salinity minimum tend to be $26.8\sigma_\theta$?" has been an important question which scientists have been asking since Reid (1965) presented his paper on the formation process of North Pacific Intermediate Water. No satisfactory answer has been found yet.

Meanwhile the general circulation field of intermediate layers in the North Pacific is obtained by applying the inverse method (Wunsch, 1978; Fukasawa et al., 1993) to ocean climatological data by Levitus (1982). The result is

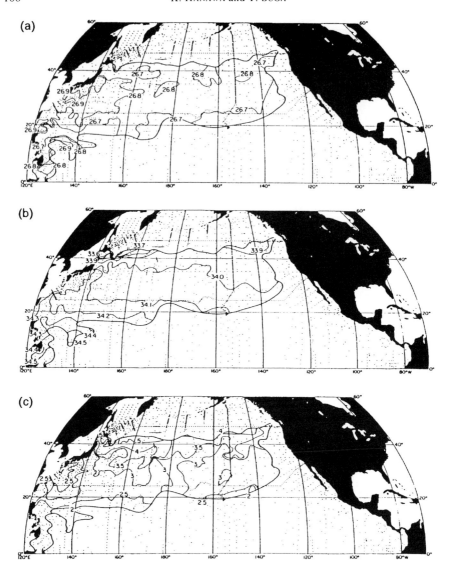

Fig. 3-24. (a) Potential density, σ_θ, (b) salinity, and (c) oxygen (ml·l^{-1}) of North Pacific
Intermediate Water, defined as all salinity minima in the density range of 26.6 to 27.0σ_θ,
excluding salinity minima with very low oxygen near the eastern boundary (Talley,
1993).

consistent with the above discussion in that the intermediate water mass is
supplied from the Okhotsk Sea and an anticyclonic circulation is dominant in the
subtropics (Fig. 3-25). However, the circulation field suggests an existence of
more than one flow core associated with several circulation cells. How does such

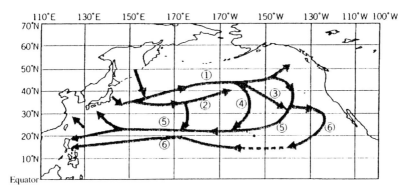

Fig. 3-25. Flow pattern of the intermediate layer including North Pacific Intermediate Water, based on the result by inverse calculation using Levitus' (1982) data (Fukasawa, 1992).

a complex circulation path of North Pacific Intermediate Water reflect the process in which information entering from the surface spreads into the intermediate layer? This is another interesting problem waiting to be solved.

REFERENCES

Bingham, F. M. (1992): *J. Geophys. Res.*, **97**, 11177–11189.
Bingham, F. M. *et al.* (1992): *J. Oceanogr.*, **48**, 405–425.
Bryden, H. L. (1973): *Deep-Sea Res.*, **20**, 401–408.
Cannon, G. A. (1966): *Deep-Sea Res.*, **13**, 1139–1148.
Cessi, P. (1988): *J. Phys. Oceanogr.*, **18**, 662–682.
Craig, P. D. and M. L. Banner (1994): *J. Phys. Oceanogr.*, **24**, 2546–2559.
Craik, A. D. D. and S. Leibovich (1976): *J. Fluid Mech.*, **73**, 401–426.
Denman, K. L. (1973): *J. Phys. Oceanogr.*, **3**, 173–184.
De Szoeke, R. A. (1980): *J. Phys. Oceanogr.*, **10**, 1439–1455.
Ekman, V. W. (1905): *Ark. Met. Astr. Fys.*, **2**, 1–52.
Emery, W. J. and J. S. Dewar (1982): *Prog. Oceanogr.*, **11**, 219–305.
Fox, C. J. J. (1909): *Trans. Faraday Soc.*, **5**, 68–87.
Fukasawa, T. (1992): *Kagaku*, **62**, 616–624 (in Japanese).
Fukasawa, T. *et al.* (1993): *Gekkan Kaiyo Kagaku, Gogai*, No. 4, 56–62 (in Japanese).
Garwood, R. W., Jr. (1979): *Rev. Geophys. Space Phys.*, **17**, 1507–1524.
Halkin, D. and T. Rossby (1985): *J. Phys. Oceanogr.*, **15**, 1439–1452.
Hanawa, K. and I. Hoshino (1988): *J. Mar. Res.*, **46**, 683–700.
Hanawa, K. and Y. Toba (1981): *Tohoku Geophys. J.*, **28**, 161–173.
Hasunuma, K. (1978): *Bulletin of the Ocean Research Institute, University of Tokyo*, **9**, 47.
Hellerman, S. and M. Rosenstein (1983): *J. Phys. Oceanogr.*, **13**, 1093–1104.
Jenkins, W. J. (1980): *J. Mar. Res.*, **38**, 533–569.
Kawabe, M. (1985): *J. Oceanogr. Soc. Japan*, **41**, 307–326.
Kawabe, M. and K. Kawasaki (1993): *JODC Manual Guide No. 4*, Japan Oceanographic Data Center, 68 pp. (in Japanese).
Keffer, T. (1985): *J. Phys. Oceanogr.*, **15**, 509–523.
Kraus, E. B. and J. S. Turner (1967): *Tellus*, **19**, 98–106.
Levitus, S. (1982): *NOAA Prof. Paper 13*, 173 pp.

Lewis, E. L. and N. P. Fofonoff (1979): *J. Phys. Oceanogr.*, **9**, 446.
Lewis, E. L. and R. G. Perkin (1978): *J. Geophys. Res.*, **83**, 466–478.
Lewis, E. L. and R. G. Perkin (1981): *Deep-Sea Res.*, **28A**, 307–328.
Luyten, J. R. *et al.* (1983): *J. Phys. Oceanogr.*, **13**, 292–309.
Lynn, R. J. and J. L. Reid (1968): *Deep-Sea Res.*, **15**, 577–598.
Madsen, O. S. (1977): *J. Phys. Oceanogr.*, **7**, 248–255.
Marshall, J. C. *et al.* (1993): *J. Phys. Oceanogr.*, **23**, 1315–1329.
Masuda, A. (1981): *Gekkan Kaiyo Kagaku*, **13**, 487–494 (in Japanese).
Masuzawa, J. (1969): *Deep-Sea Res.*, **16**, 463–472.
Masuzawa, J. and K. Hasunuma (1977): *Kaiyo Kagaku Kiso Koza 4, Physical Oceanography IV*, Tokai University Press, Tokyo, pp. 1–114 (in Japanese).
Masuzawa, J. and K. Nagasaka (1975): *J. Mar. Res.*, **33**(Suppl.), 109–116.
McCartney, M. S. and L. D. Talley (1982): *J. Phys. Oceanogr.*, **12**, 1169–1188.
McDougall, T. J. (1987): *J. Phys. Oceanogr.*, **17**, 1950–1964.
Mellor, G. L. and P. A. Durbin (1975): *J. Phys. Oceanogr.*, **5**, 718–728.
Mellor, G. L. and T. Yamada (1974): *J. Atmos. Sci.*, **31**, 1791–1806.
Millero, F. J. *et al.* (1980): *Deep-Sea Res.*, **27A**, 255–264.
Montgomery, R. B. (1937): *Bull. Amer. Meteor. Soc.*, **18**, 210–212.
Montgomery, R. B. (1958): *Deep-Sea Res.*, **5**, 134–148.
Montgomery, R. B. and E. D. Stroup (1962): Johns Hopkins Oceanogr. Studies, No. 1, 68 pp.
Montgomery, R. B. and W. S. Wooster (1954): *Deep-Sea Res.*, **2**, 63–70.
Nakamura, H. (1996): *J. Oceanogr.*, **52**, 171–188.
Niiler, P. P. (1975): *J. Mar. Res.*, **33**, 405–422.
Niiler, P. P. and E. B. Kraus (1975): *Modeling and Prediction of the Upper Layers of the Ocean*, ed. by E. B. Kraus, Pergamon, pp. 143–172.
Open University (1989): *Ocean Circulation*, Pergamon Press, 238 pp.
Pickard, G. L. and W. J. Emery (1990): *Descriptive Physical Oceanography, An Introduction*, fifth enlarged edition, Pergamon Press, 320 pp.
Price, J. F. (1981): *J. Phys. Oceanogr.*, **11**, 153–175.
Reid, J. L. (1965): *Johns Hopkins Oceanogr. Studies*, No. 2, 85 pp.
Reid, J. L. (1973): *Deep-Sea Res.*, **20**, 51–68.
Reid, J. L. and R. J. Lynn (1971): *Deep-Sea Res.*, **18**, 1063–1088.
Roemmich, D. and B. Cornuelle (1992): *J. Phys. Oceanogr.*, **22**, 1178–1187.
Stommel, H. (1979): *Proc. Nat. Acad. Sci. U.S.A.*, **76**, 3051–3055.
Suga, T. and K. Hanawa (1990): *J. Mar. Res.*, **48**, 543–566.
Suga, T. and K. Hanawa (1995a): *J. Phys. Oceanogr.*, **25**, 958–970.
Suga, T. and K. Hanawa (1995b): *J. Phys. Oceanogr.*, **25**, 1012–1017.
Suga, T. *et al.* (1989): *J. Phys. Oceanogr.*, **19**, 1605–1618.
Suga, T. *et al.* (1997): *J. Phys. Oceanogr.*, **27**, 141–152.
Sverdrup, H. U. *et al.* (1942): *The Oceans, Their Physics, Chemistry and General Biology*, Prentice-Hall, 1087 pp.
Talley, L. D. (1985): *J. Phys. Oceanogr.*, **15**, 633–649.
Talley, L. D. (1988): *J. Phys. Oceanogr.*, **18**, 89–106.
Talley, L. D. (1991): *Deep-Sea Res.*, **38**(Suppl.), S171–S190.
Talley, L. D. (1993): *J. Phys. Oceanogr.*, **23**, 517–537.
Tanimoto, Y. *et al.* (1993): *J. Clim.*, **6**, 1153–1160.
Tatsumi, T. and K. Goto (1976): *Stability Theory of Flow*, Sangyo Tosho, 275 pp. (in Japanese).
Toba, Y. and H. Kawamura (1995): *Air-Water Gas Transfer*, ed. by B. Jähne and E. Manahan, AEON Verlag, Heidelberg, pp. 1–8.
Tsuchiya, M. (1970): *Kaiyo Kagaku Kiso Koza 10, Science of Sea Water*, Tokai University Press, Tokyo, pp. 141–175 (in Japanese).
Tsuchiya, M. (1982): *J. Mar. Res.*, **40**(Suppl.), 777–799.
Tsunogai, S. and S. Noriki (1983): *Kaiyo Kagaku*, ed. by M. Nishimura, Sangyo Tosho, 286 pp. (in Japanese).

Uda, M. (1935a): *Chuo Suishihou*, **6**, 1–130 (in Japanese).

Uda, M. (1935b): *Umi to Sora*, **15**, 445–452 (in Japanese).

UNESCO (1981a): UNESCO Technical Papers in Marine Science, No. 37, 144 pp.

UNESCO (1981b): UNESCO Technical Papers in Marine Science, No. 38, 192 pp.

UNESCO (1983): UNESCO Technical Papers in Marine Science, No. 44, 53 pp.

Weiss, R. F. (1970): *Deep-Sea Res.*, **17**, 721–735.

Woods, J. D. (1980): *Quart. J. Roy. Meteor. Soc.*, **106**, 379–394.

Worthington, L. V. (1959): *Deep-Sea Res.*, **5**, 297–305.

Wunsch, C. (1978): *Review of Geophys. and Space Phys.*, **16**, 538–620.

Wüst, G. (1929): Veröff. Inst. Meeresk. Univ. Berlin, neue Folge, A, 20, 64 pp.

Wüst, G. (1935): *Wiss. Ergeb. Dtsch. Atl. Exped.*, **6**, 109–288.

Yuan, X. and L. D. Talley (1992): *J. Phys. Oceanogr.*, **22**, 1302–1316.

Zhang, H. M. and N. G. Hogg (1992): *J. Mar. Res.*, **50**, 385–420.

Chapter 4

Large Scale Ocean–Atmosphere Interactions

Kimio HANAWA

Ocean–Atmosphere Interactions, Ed. Y. Toba, pp. 111–142.
© by TERRAPUB / Kluwer, 2003.

Large Scale Ocean–Atmosphere Interactions

Kimio HANAWA

The wind goeth toward the south, and turneth about
unto the north; it whirleth about continually, and the
wind returneth again according to his circuits.

—Ecclesiastes 1:6

4.1 TIME SCALE OF CLIMATE CHANGE AND THE ROLE OF THE OCEAN

Climate is a realization of an averaged state of an atmosphere, but five spheres, namely, atmosphere, hydrosphere, geosphere, biosphere, and cryosphere all participate in the formation of climate. These five spheres interact with and depend upon each other in a very complex manner. In this sense, they can be generically called a "climate system". Each one of this system's constituents has its own characteristics so that their time scales differ from each other. Table 4-1 summarizes the characteristic time scales of various factors that cause climate change (Kutzbach, 1976). In this table the ocean–atmosphere interactions are associated with the time scales from seasons to several 1000 years.

As will be discussed in the following section in detail, the ocean stores heat 1000 times more than the atmosphere; also the relaxation time of the ocean for changes at the boundary is far longer than in the atmosphere. As a result, the ocean plays a major role in the climate variation of time scales from as short as seasons to as long as 1000 years. In this chapter, a general review of the characteristics of the ocean's role will be given, and then recent developments of studies on variations in the ocean–atmosphere-coupled system will be introduced, focusing on time scales from a few years to decades.

4.2 OCEAN–ATMOSPHERE INTERACTION SYSTEM

4.2.1 What is the ocean–atmosphere interaction system?

Let us take the ocean–atmosphere system as a simplified climate system. The concept of the variations of this system is schematically drawn in Fig. 4-1. Most of the radiation energy from the sun, which consists mainly of visible light (solar radiation), passes through the air and reaches the ocean surface, which occupies about 70% of the earth's surface. In other words, the atmosphere is "transparent" with regards to solar radiation, but the ocean is not. The energy that has reached the ocean surface is absorbed by a very thin surface layer. Because

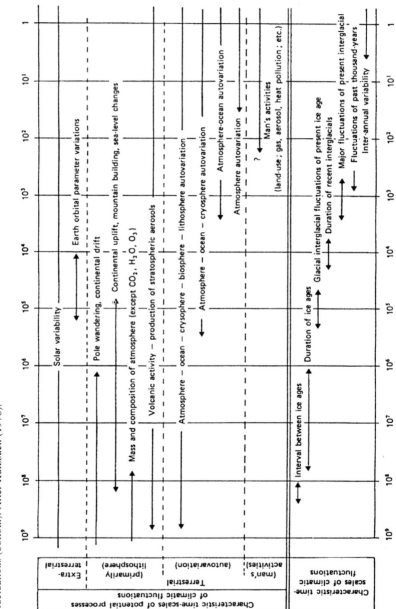

Table 4-1. Examples of potential processes involved in climate fluctuations (top) and characteristic time scales of observed climatic fluctuations (bottom). After Kutzbach (1976).

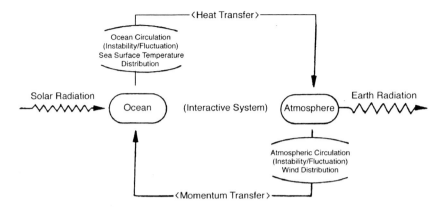

Fig. 4-1. Schematic picture of atmosphere–ocean interactive system.

the earth has a spherical shape, and it rotates around the sun with a tilting axis of rotation, the heat quantity the ocean receives from the sun varies depending on latitudes and seasons. As a result, the sea surface temperature is distributed non-uniformly: warmer in low latitudes and colder in high latitudes on the large scale. Consequently, the density of seawater in the surface layer is smaller in low latitudes and larger in high latitudes. Also, the thermal energy the atmosphere receives from the ocean is uneven in space and a similar pattern of the distribution of density is formed in the atmosphere. This distribution of density generates convective motion both in the atmosphere and the ocean.

The convection of the atmosphere on the large scale is organized as a systematic motion over the earth, influenced by a complex combination of its density stratification and the Coriolis force due to the rotation of the earth. The motion of the atmosphere is "wind". The wind blowing over the ocean surface provides momentum to seawater, by dragging the ocean surface. That is, the wind stress acts on the sea surface, the surface layer setting in motion. The large-scale oceanic motion also becomes systematic under the combined influence of the Coriolis force, density stratification and the continental boundaries in the north-south direction. The motion of the ocean redistributes the internally stored heat, and changes the distribution of sea surface temperatures, which in turn generates a new response of the atmosphere.

Both atmosphere and ocean are strongly non-linear systems with a large degree of freedom, so that in spite of the fact that the incoming energy (solar radiation) is constant, their motions involve "fluctuations (instabilities)". These fluctuations work as external forces to each other. That is, the ocean–atmosphere coupled system cannot be in a stationary state, but constantly oscillates around an equilibrium point. Further, it is known that sometimes transition from one state to a completely different state occurs (two or more equilibrium points are possible). The atmosphere and the ocean constitute a mutually interacting system with feedback circuits.

4.2.2 Heat storage of the ocean and relaxation time

The importance of the ocean's role can be summarized in the following two points. One is that the ocean has a heat storage of 1000 times greater than the atmosphere does. The other is that the mechanism of stratification differs in the two fluids: the atmosphere is of "radiation-convection process" type, whereas the ocean is of "advection-diffusion process" type, which requires a much longer processing time. In addition to these points, another very important aspect of the ocean for the formation and the change of climate is that seawater, which occupies 97% of water on the earth's surface, has the capacity to dissolve a large amount of various substances such as CO_2. But this topic is beyond the scope of this chapter and will not be discussed here.

Let us estimate a ratio of the heat stored in the atmosphere and that in the ocean by a simple calculation. The quantity of heat storage can be obtained by integrating over the entire fluid using the values of fluid density, specific heat at constant pressure and temperature. The approximate ratio estimated is 1:1000. Namely the ocean stores heat 1000 times greater than the atmosphere does. That is why we can say "an upper layer of a few meters has the same heat storage as the entire atmosphere".

Next we will discuss how the ocean and the atmosphere are stratified. The stratification of the atmosphere is determined by a "radiation-convection process" as mentioned before. Materials emitted from the lower boundary (near the earth's surface) spreads quickly into the troposphere. It takes them from a few days to two weeks to go around the earth in the same latitude. The atmosphere receives thermal energy from the lower boundary while emitting it out of the earth by long-wave radiation. Stratification is basically stable, although the heating from the lower boundary can easily destroy the stability. The atmosphere contains a large amount of vapor which can release latent heat at condensation, so that movement of vapor can be regarded as movement of heat. This internal heating makes the atmosphere unstable and effectively destroys the stratification. The non-uniform stratification produced by the convection process is made uniform rapidly by an infrared radiation process. For this reason, the adjustment time of the earth's atmosphere to variations of boundary conditions is estimated to be a month or so. If we call this adjustment time "relaxation time", we can say that the atmosphere's relaxation time is of the order of one month. People, living in a mid latitude, can enjoy four seasons, because the atmosphere has a relaxation time much shorter than one year, quickly responding to the seasonal change of solar radiation or the lower boundary condition.

Stratification in the ocean is formed by an "advection-diffusion process". Analyzing chemical tracers, we can estimate the time that has elapsed since seawater's contact to the atmosphere. The result shows that seawater in the intermediate layer of the North Pacific is the oldest in the world's oceans: the elapsed time is estimated as 2000 years (Tsunogai, 1981). It is to be noted that the internal motion of the ocean is extremely slow. Unlike the atmosphere the ocean does not have an internal source of heat and the geothermal heat flow from the sea floor is very small. Hence, stratification is determined by heating and cooling and the difference between precipitation and condensation at the upper boundary. The absorption coefficient of seawater for solar radiation (visible light) is very large and in almost all oceans, almost all of the thermal energy is absorbed within the upper 100 meters. Consequently, when the ocean receives heat, it forms stable stratification, whereas when the ocean loses heat, convection occurs, destroying stratification. Heat is transported into deep ocean by turbulent diffusion and advection processes. "The turbulent diffusion time", calculated as the squared spatial scale divided by the vertical diffusion coefficient, turns out to be of the order of 1000 years, assuming a vertical scale of 4000 meters and a vertical eddy diffusion coefficient of 1 cm^2/s. This time scale can be regarded as that of the ocean's response to the variation of thermal conditions at the upper boundary. To sum up, both the temporal scale of advection and that of diffusion are of the order of 1000 years, and the relaxation time of the ocean is of the order of 1000 years. Therefore, the layer that is subject to seasonal change is restricted to the surface of the ocean from a few tens of meters to 100 meters. For instance, it can be estimated that in the ocean with a vertical eddy diffusion coefficient of 1 cm^2/s, information reaches to the depth of about 60 m. That is, there are no seasons in the majority of the oceans.

The ocean is a system that has greater "thermal inertia" than the atmosphere, for the two reasons discussed above; it has much larger heat storage and longer relaxation time. This is clearly shown in the following example. Figure 4-2 illustrates the result of the experiment on the sudden increase and decrease of carbon dioxide using an ocean–atmosphere-coupled model by Manabe et al. (1991). The figure shows how the mean temperature on the earth changes if the carbon dioxide increases or decreases by 1% per year. In either case, the temperature change in the Southern Hemisphere is slower than in the Northern Hemisphere. This is because the area of the continent in the Southern Hemisphere is only half of that in the Northern Hemisphere, the ocean covering 81% of the surface, and its large thermal inertia keeps the temperature changes smaller.

In numerical models for forecasting short-term weather, oceanic motion is not taken into consideration because of the large thermal inertia: the sea surface temperature values given initially are used throughout the whole integration. In the models, although thermal flux and momentum flux between the ocean and the atmosphere are depicted, the sea surface temperatures do not change. That is to say, the ocean behaves as if it were a heat source or a heat sink of infinite capacity. In this case, the ocean's large thermal inertia allows the sea surface temperatures to be assumed constant for a short period, even if there is some heat flux across the sea surface. As a matter of course, if the models involve integration for a few months or longer, they need to forecast sea surface temperatures, too.

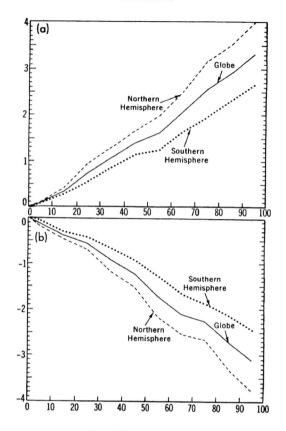

Fig. 4-2. Temporal variations of the differences in area-averaged, decadal-mean surface air temperature (°C) between integrations: (a) integration under growing CO_2 concentration with the rate of 1%/year and integration with steady concentration, and (b) integration under decreasing CO_2 concentration with the rate of 1%/year and integration with steady concentration. Solid, dashed and dotted lines indicate the differences over the globe, and Northern and Southern Hemispheres, respectively. After Manabe *et al.* (1991).

4.3 FLUXES BETWEEN THE ATMOSPHERE AND THE OCEAN

In the previous section the sea surface temperature was given as the lower boundary condition to the atmosphere, but strictly speaking, substances that are really exchanged between the ocean and the atmosphere are fluxes of heat, momentum (wind stress), and fresh water through the boundary between the two fluids. In order to gain proper and quantitative understanding of interactions between the ocean and the atmosphere, accurate evaluation of the various fluxes is necessary. In this section, practical problems of estimating the fluxes will be discussed.

4.3.1 Importance of evaluation of sea surface fluxes

In recent research programs for understanding the climate change mechanism, e.g., TOGA (Tropical Ocean and Global Atmosphere Programme) and WOCE (World Ocean Circulation Experiment), accurately evaluating the fluxes of heat, momentum and fresh water across the sea surface as time series over the entire globe is recognized as being quite important. In fact for the past 15 years, numerous researchers have been engaged in evaluating these fluxes in the world's oceans.

One of the reasons why these fluxes have been studied by so many researchers is that large quantities of maritime meteorological observations reported by ships have been accumulated and stored in computers in an easily accessible form. For instance, now widely used COADS (Comprehensive Ocean–Atmosphere Data Set) are made for the purpose of including all maritime weather information since the 1850s. Another reason is that direct measurements of the fluxes of the sea have increased and the bulk coefficient (see the next section) is believed to be more reliable than before.

Meanwhile, as high-speed computers have been developed, numerical models as a tool for experiments or simulations and general circulation models which combine these have been exploited, and they require the flux values. In other words, the fluxes of the sea surface represent the boundary conditions that drive the oceanic and atmospheric models. Also since, for coupled ocean–atmosphere models, the fluxes are physical quantities that represent the intensity of interactions, they can be a subject of comparison and examination to see whether or not the coupling of the ocean and the atmosphere is properly done. Furthermore, by globally estimating fluxes, we can evaluate the meridional transport of heat and fresh water carried by seawater. This is called the surface flux method. These meridional transports also provide a "target" that should be reproduced by a correct climate model.

4.3.2 Evaluation of sea surface fluxes by the bulk method

The method of evaluating sea surface fluxes is divided into two categories. One is the direct method, the other the indirect method. The former includes the eddy correlation method for sensible (latent) heat in which temporal variations of vertical velocity and air temperature (moisture) are measured and their correlation calculated, and the dissipation method for momentum in which very short-period variations are measured. In the latter, since the measuring device is within the atmospheric boundary layer, we make use of the fact that the averaged physical quantities are distributed logarithmically. This includes the gradient method in which detailed distribution in the vertical direction is examined, and the bulk method, which is a simplification of the gradient method. The bulk method is most appropriate for obtaining long-term fluxes in a vast area of the sea that is suitable for studying large-scale ocean–atmosphere interactions.

In the bulk method, fluxes of momentum (wind stress), sensible heat and latent heat are evaluated by using quantities averaged over a certain period of time at the sea surface and at a certain height, usually at 10 m height above the surface. They are given by the following formulae.

$$\tau_x = \rho C_D |V| u \qquad\qquad \text{: East-west component of momentum} \qquad (4\text{-}1)$$

$$\tau_y = \rho C_D |V| v \qquad\qquad \text{: North-south component of momentum} \qquad (4\text{-}2)$$

$$Q_H = \rho C_p C_H |V|(T_s - T_a) \quad \text{: Sensible heat} \qquad (4\text{-}3)$$

$$Q_E = \rho L C_E |V|(q_s - q_a) \qquad \text{: Latent heat} \qquad (4\text{-}4)$$

where ρ, C_p, and L are the air density, specific heat at constant pressure and latent heat by evaporation, respectively. $|V|$ is the absolute value of wind velocity, u and v are its east-west and north-south components, T_a the air temperature, q_a the air specific humidity, T_s the sea surface temperature and q_s the specific humidity close to the surface, assumed to be saturated. The sea surface temperature is the water temperature averaged from the sea surface to a few tens of centimeters, namely the bulk water temperature. The rationale for the bulk coefficients was fully discussed in Chapter 2, but here it is worth mentioning that Eqs. (4-1) and (4-2) are components of Eq. (2-116), and Eqs. (4-3) and (4-4) correspond to Eqs. (2-117) and (2-118).

C_D, C_H, and C_E are called the exchange coefficients or bulk coefficients for momentum, sensible heat, and latent heat, all having dimensionless values. They are functions of wind speed and atmospheric boundary layer stability, and of the order of 10^{-3}. The stability is also a function of wind speed and the difference between water temperature and air temperature.

Most of the bulk coefficients presently used are parameterized as coefficients for the data averaged over ten minutes or so. This is based on the fact that the spectra of physical quantities within the atmospheric boundary layer have peaks below several minutes of periods, and have little energy between 10 minutes and several hours of variation. That is, the present bulk coefficients are parameterization of these high-energy parts below several minutes. Hence, in the bulk method the above-mentioned fluxes can be estimated as long as wind speed, wind direction, seawater temperature and dew-point temperature are known. They are part of maritime meteorological data, which ocean liners report, together with cloudiness and pressure data.

Empirical formulas are also available for estimating the remaining two elements of thermal radiation, e.g., shortwave and long-wave radiation from maritime meteorological information. They are not generally called the bulk method, but in a broad sense they can be included in the bulk method. Although numerous researchers proposed a variety of empirical formulas, typical examples are given below.

$$Q_S = (1 - \alpha)Q_{S0}(1 - AC) \qquad \text{: Net shortwave irradiation} \qquad (4\text{-}5)$$

$$Q_B = \varepsilon_\sigma T_S^{\,4}\left(a - be^{1/2}\right)(1 - BC) \quad \text{: Net long-wave irradiation} \qquad (4\text{-}6)$$

where, α, ε, and σ are the albedo at the sea surface, the emission rate of seawater, and the Stefan–Boltzmann constant, respectively. Q_{S0}, C, and e are shortwave radiation at the top of the atmosphere (uniquely determined by the time and the position), cloudiness represented by a number from 0 to 10, and the vapor pressure in the air, respectively. All A, B, a, and b are constants determined empirically.

4.3.3 Problems in evaluating sea surface fluxes

It is very effective to study variations of fluxes at the sea surface for understanding air–sea interactions. However, it should be borne in mind that there are still many problems to be considered on flux estimation. For illustrating these problems, let us set the following situation. "We will estimate the monthly averaged fluxes at the sea surface in a vast area in the ocean at grids of a few degrees of latitude and longitude for a long period of time. The necessary data will be obtained from an existing data file." Under this situation, possible errors that may occur will be; (1) errors included in measurements, (2) errors in creating the data base, (3) uncertainty of the bulk coefficients and the empirical formulas, (4) errors from inappropriate use of the bulk formulas, and (5) errors due to biased sampling of data. All of these errors are quite difficult to deal with, and very careful attention should be paid in evaluating the fluxes.

For example, yearly averaged estimates by different investigators of the heat flux over the "warm pool" in the western equatorial Pacific, a key area for ENSO (El Niño/Southern Oscillation) events, differs significantly, often by 20 to 100 W m^{-2}. A difference of 80 W m^{-2} between the maximum and minimum estimates, when converted to the temperature change of the seawater column in the upper 100 m layer, would produce a difference of 6°C in a year. It is needless to say that this difference is beyond the permissible limit.

In order to increase the reliability of estimation in the framework of the bulk method, efforts to remove various uncertainties for each category of the possible errors listed above should be done. For example, the following points can be noted.

(1) Data quality control

It is necessary to improve the quality of data to be used. The methods and conditions of measurements have sometimes changed. For instance, several corrections should be properly made: corrections for air temperature measurements made by sailing boats and those made by steel ships, corrections for the difference of sea surface temperatures from a bucket sampling and an intake sampling, and corrections for converting from Beaufort scales to wind velocities, etc.

(2) Data rescue and archive

There are some data files in the world not stored in a computer. It is necessary to add them to a single database.

(3) Enhancement of observation

The amount of marine meteorological data presently reported is not sufficiently large and it must be increased. It is necessary to make an effort to ask all available ships to collect and share marine meteorological information. Also installation of meteorological sensors to drifting buoys and moored buoys, which are already used in the TOGA and WOCE programs should be promoted.

(4) Improvement of bulk coefficients and empirical formulas

Bulk coefficients for momentum, latent heat, and sensible heat fluxes as well as the empirical formulae for the radiation flux should be more precisely determined. These problems in estimating fluxes by the bulk method were documented in detail by Hanawa (1993).

So far the bulk method has been the only way of estimating sea surface fluxes, but recent progress in remote sensing techniques using satellites have made it possible to evaluate fluxes using satellite data.

Also as numerical models have been improved, some of the fluxes can be estimated by these models (e.g., Ishii *et al.*, 1994). This aspect will be discussed in detail in Chapter 7.

4.3.4 Distribution of surface fluxes

In this section, typical distributions of sea surface fluxes will be shown. They are all estimated using marine meteorological observations reported by ships. Here, note that they are not chosen from the viewpoint that they are the most accurately estimated among various data sets. Since they are based on data collected from ship observations, ice covered regions and some areas in the southern hemisphere often appear as large gaps. Therefore, it is not possible to calibrate the heat flux or fresh water flux schemes by achieving a vanishing meridional heat or freshwater transport at the Antarctica and the North Pole.

a) Wind stress

The wind stress data most widely used today for driving numerical models of the ocean were made by Hellerman and Rosenstein (1983). Approximately 35 million pieces of data collected from ship reports are used for the calculations. Figures 4-3(a), (b) and (c) show yearly mean values of wind stress in January, February and July. Comparison of the three figures indicates that strong winds constantly blow over the Antarctic Circumpolar Sea. Strong southeast trade winds also blow throughout the year in the Southern Hemisphere of the Indian Ocean, but in the Northern Hemisphere the wind directions change drastically between January and July. This is what we call Asian Monsoon. Particularly in July, the southeast winds from the African continent to the offshore of the Arabian Peninsula exceed 0.3 N m^{-2}. It is seen that both in the North Pacific and the North Atlantic in January, the strong burst comes out of the continents, the (a)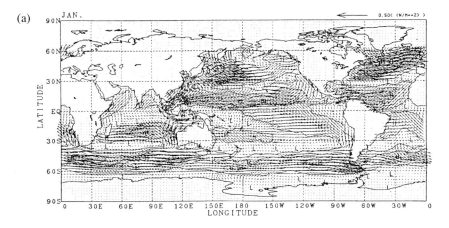

(b)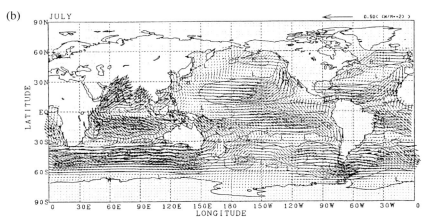

(c)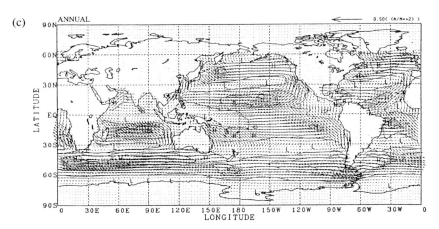

Fig. 4-3. Climatologies of wind stress vectors. (a) January, (b) July and (c) annual mean field. Original data are those of Hellerman and Rosenstein (1983). These figures were prepared by Mr. Ikuo Yoshikawa, Japan Meteorological Agency.

wind speed surpassing 0.25 N m^{-2} in some areas. In July, an anti-cyclonic circulation appears as subtropical high pressure is developed.

b) Heat flux

Oberhuber (1988) obtained climate values of the heat flux at the sea surface for 30 years from 1950 to 1979, using the aforementioned COADS data. Figures 4-4 to 4-7 show distributions of yearly mean values of shortwave radiation, long-wave radiation, sensible heat and latent heat fluxes. The panel on the right in each figure represents the distribution averaged in the east-west direction, latitude by latitude. A positive value indicates downward flux.

Shortwave radiation (Fig. 4-4) is essentially a function of cloud cover and daily averaged sun elevation. It exceeds 200 W m^{-2} in the low latitudes in the South Pacific, the South Atlantic, the western part of the Indian Ocean, and the northeast of Australia. It decreases towards both Poles: less than 100 W m^{-2} north of 45°N. The east-west average has a minimum near 10°N. This narrow region, reflects reduced solar heat flux due to the existence of ITCZ, and the subtropics higher values result from less cloudiness.

Long-wave radiation (Fig. 4-5) is influenced by the atmospheric water vapor content and the cloud cover. Consequently, we find lower heat loss in lower latitudes due to the high humidity and higher heat loss in the subtropics, which has typically dry and cloud free air. The range of heat loss is between −30 and −70 W m^{-2}, having slightly larger values in the western part in mid latitudes.

The sensible heat flux (Fig. 4-6) is generally small because the sea surface

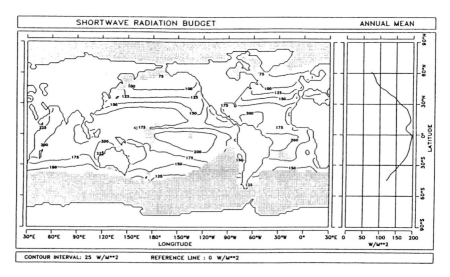

Fig. 4-4. Annual mean net shortwave radiation. Units in W/m^2. Positive values mean downward: the ocean gains heat. Right panel shows latitudinal distribution of zonal average. After Oberhuber (1988).

and the air temperatures do not differ significantly, lying between 0 and −10 W m^{-2} in almost all areas, with exceptions in the Kuroshio and the Gulf Stream regions where it exceeds −50 W m^{-2}. This is because the Kuroshio and the Gulf Stream transport warm water from low latitudes and cold, dry air of the continental

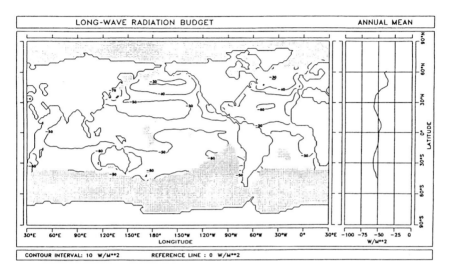

Fig. 4-5. As in Fig. 4-4 except for annual mean long-wave radiation. After Oberhuber (1988).

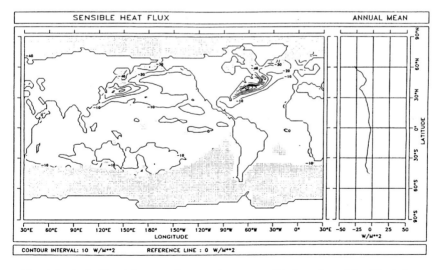

Fig. 4-6. As in Fig. 4-4 except for annual mean sensible heat flux. After Oberhuber (1988).

origin flows over it, making a big difference between air and sea surface temperatures. Small values in the East Australian Current region are probably due to the small size of the Australian continent, which cannot generate cold, dry winds.

The latent heat flux (Fig. 4-7) is controlled by wind speed, air humidity and temperature. Therefore, we obtain systematically higher values at lower latitudes due to the higher temperature and resultant higher absolute humidity there. It is considerably larger than sensible heat, between -100 W m^{-2} and -200 W m^{-2} in the low and mid latitudes, decreasing towards the poles. It takes large values in the Kuroshio and Gulf Stream regions east of the continents, as was the case for sensible heat. It is also large in subtropical high-pressure zones outside of the equator. It is less than 100 W m^{-2} in the upwelling region in the eastern Pacific.

Out of the four fluxes shown above, shortwave radiation and latent heat are predominant. The long-wave radiation contributes a heat loss of typically -50 W m^{-2}. The sensible heat flux is negligible except for the area of the western boundary currents. Figure 4-8 shows the annual mean of the net heat flux. As is expected, areas of strong heat gain exist at the equator, particularly in the cool upwelling region in the eastern Pacific where the value exceeds 100 W m^{-2}. Part of the high-latitude ocean gains heat, too. A strong heat loss exceeding -150 W m^{-2} appears in regions of the western boundary currents in mid latitudes.

c) Freshwater flux

The global distribution of the net downward flux of fresh water is shown in Fig. 4-9 according to Oberhuber (1988). It is the distribution of climatological values of the difference $(P - E)$ of precipitation P and evaporation E. Evaporation

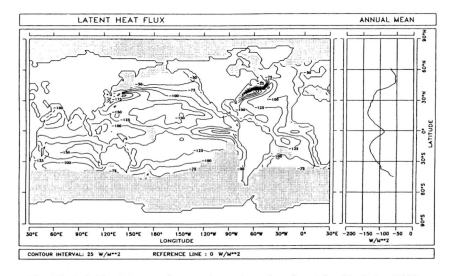

Fig. 4-7. As in Fig. 4-4 except for annual mean latent heat flux. After Oberhuber (1988).

can be simply derived in the process of calculating latent heat (it is actually the latent heat Q_E divided by the latent heat L of the evaporation). But since very few measured values for precipitation in the ocean are available, measurements of precipitation for continents and islands with estimated values using satellites are

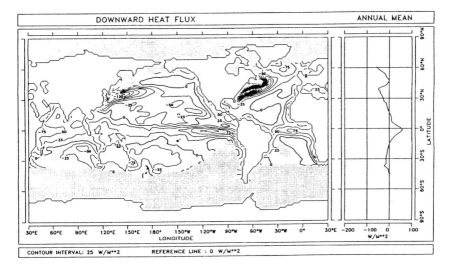

Fig. 4-8. As in Fig. 4-4 except for annual mean net heat flux. After Oberhuber (1988).

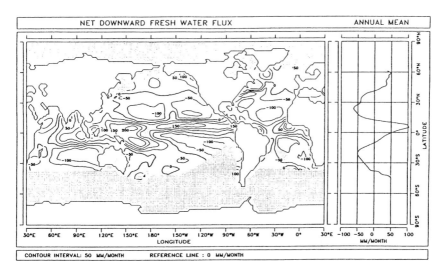

Fig. 4-9. As in Fig. 4-4 except for annual mean fresh water flux (*P–E*: precipitation minus evaporation). Positive values mean downward: the ocean gains freshwater. Units in mm/month.

used (Shea, 1986).

There is more precipitation than evaporation in the subtropical zonal belt. In the ocean north of Australia, precipitation exceeds 2 m/year. On the contrary, excess evaporation occurs in mid latitudes where subtropical high pressure develops.

4.4 TELECONNECTION AND CLIMATE IN JAPAN

It has been discovered recently that, as a fluctuating phenomenon of the atmosphere, a large-scale circulation field varies in an organized pattern, in addition to the synoptic-scale phenomenon such as moving high and low pressures. This phenomenon is a teleconnection pattern whose substance is considered to be a standing Rossby wave train. The teleconnection pattern is considered to be a realization of the ocean–atmosphere interactions and is being actively studied today.

4.4.1 Teleconnection

Meteorological parameters at widely separated points on earth that are seemingly not related may synchronize to fluctuate. This is called "teleconnection". As modern meteorological observation nets were extended all over the world and the data were accumulated, some climate patterns were known to change, synchronizing on a very large scale. The Southern Oscillation (Walker and Bliss, 1932) was one example. In the early 1980s, Wallace and his collaborators at the University of Washington revealed the existence of teleconnection and clarified its significance to meteorology and climate variation.

Wallace and Gutzler (1981), using long time series of data from the 500 hPa height field, calculated temporal correlation coefficients between all possible pairs of grid points to study their distribution. Figure 4-10 is an example of their results, showing one-point correlation with reference to a grid point located at (45°N, 165°W) and at a height of 500 hPa (5000–6000 m above the ground). Positive values are shown in the vicinity of the reference point and on the east coast of the North American continent, while negative values are in the central Pacific in low latitudes and in the western coast of the North American continent. Other than those areas no systematic correlation is seen. This pressure series, comprising the four centers, ranging from the equatorial Pacific to mid latitudes to Alaska to the eastern coast of America, was named PNA (Pacific/North American) pattern. This is interpreted as a standing Rossby wave generated by a heat source from the equatorial region (Horel and Wallace, 1981). The pattern shown in the correlation coefficient (Fig. 4-10) can be also confirmed by means of a composite map. If the "center of action" is defined as the best correlated place, an activity index can be calculated by a linear combination of anomalies from the mean. Figure 4-11 shows the difference between the heights of a 500 hPa surface, based on winter months with the 10 highest and 10 lowest values of

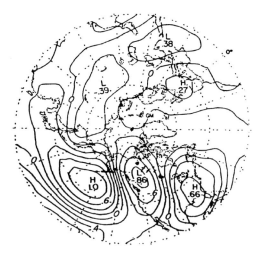

Fig. 4-10. One-point correlation map showing the correlation coefficient at 500 hPa height for the reference grid point of 45°N, 165°W. Contour interval is 0.2. After Wallace and Gutzler (1981).

Fig. 4-11. Height difference of the 500 hPa surface between two composites for strongest positive and negative values of the Pacific/North American (PNA) index. Each composite (not shown here) is constructed using 10 months out of the 45-month data set during 15 years, based on the PNA index. After Wallace and Gutzler (1981).

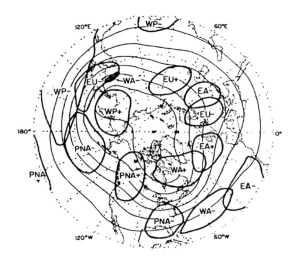

Fig. 4-12. Schematic picture of five pronounced teleconnection patterns during the Northern Hemisphere winter. ±0.6 isopleths of correlation coefficient between each of the five pattern indices and local 500 hPa height (heavy lines) are drawn on the wintertime mean 500 hPa height contours (thin lines). Contour interval is 120 m. Regions of strong correlation are labeled in terms of the respective pattern indices, and the sign of that correlation is indicated. After Wallace and Gutzler (1981).

the PNA index. At the center of action east of the central Pacific, the difference between the positive PNA index and negative PNA (anti-PNA) index amounts to 245 m in altitude.

Wallace and Gutzler found five patterns including PNA teleconnection in the variations in the wintertime Northern Hemisphere. The results are summarized in Fig. 4-12. In the figure, ±0.6 isopleths of correlation coefficient between each of the five pattern indices and local 500 hPa height (heavy lines), superimposed on wintertime mean 500 hPa height contours (lighter lines), are drawn with contour interval of 120 m. Regions of strong correlation are labeled in terms of the respective indices with which a local 500 hPa height shows the strongest correlation, and the sign of that correlation is indicated. EA is the Eastern Atlantic pattern, WA the Western Atlantic pattern, WP the Western Pacific pattern, and EU the Eurasian pattern. Since then, various teleconnection patterns have been proposed by many researchers.

Among the known teleconnection patterns the PNA pattern is considered to be most active. Although they differ by kinds of patterns and also by seasons, they usually last from a week to several tens of days. Some patterns maintain high activities for a few years or decades, in a particular season. Prediction of the teleconnection patterns will be quite useful for weather forecast. That is one of the reasons why they are an important subject of study.

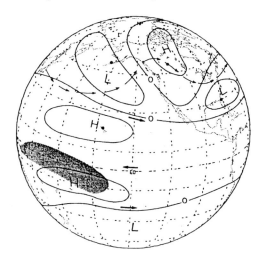

Fig. 4-13. Schematic picture of the hypothesized global pattern of middle and upper troposphere geopotential height anomalies (solid line) during a Northern Hemisphere winter which falls within an episode of warm SST in the equatorial Pacific (ENSO event). The arrows in darker type reflect the strengthening of the subtropical jets in both hemispheres along with stronger easterlies near the equator during warm episodes. The arrows in lighter type depict a mid-tropospheric streamline as distorted by the anomaly pattern. The shaded area shows regions of enhanced cirriform cloudiness and rainfall. After Horel and Wallace (1981).

4.4.2 ENSO event and climate in Japan

An ENSO event is a phenomenon due to the ocean–atmosphere interaction in the equatorial Pacific, but its influence spreads over the entire globe through teleconnection patterns. This section will review the effect of ENSO on climate in Japan. The dynamics of ENSO itself will be discussed in Chapter 5.

a) PNA pattern and wintertime climate

It is known that the PNA pattern tends to appear in winter when an ENSO event occurs. This is because heating of the atmosphere by the movement of warm water from the western equatorial Pacific to the east is likely to generate the PNA pattern. Figure 4-13 shows a schematic diagram of the pattern. As a result, low pressure appears off the west coast of the American continent, with a high pressure to its north east. This pressure pattern tends to bring about warm winter from the west coast to Alaska together with southerly winds. In the meanwhile, the pressure is reversed in the central part of the North American continent, leading to colder winter. Winter during the 1987/1988 ENSO was such a typical winter under the PNA pattern.

The appearance of the PNA pattern corresponds to the fact that the Aleutian Low moves to the east and is intensified. For this reason, the axis

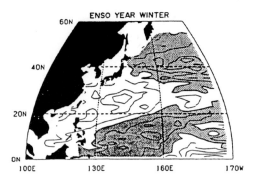

Fig. 4-14. Composite map of SST anomaly fields for the ENSO year winter (winter during the ENSO event). Units in °C. Contour interval is 0.2°C. Dotted areas show the region with negative anomalies. After Hanawa *et al.* (1988).

of the westerlies in the western part of the North Pacific moves to the north, and the strongest wind is shifted to the east. This is equivalent to the fact that the cold dry nozzle from Siberia is directed due east. Consequently, the wintertime East Asian monsoon burst is weakened near Japan. In such a year, winter in Japan tends to be warmer. Figure 4-14 shows a composite map of the wintertime sea surface temperature anomalies during an ENSO event (Hanawa *et al.*, 1988). Strong and well-ordered patterns of anomalies with positive values are seen in a zonal band along a 30°N line from the Asian coast to the International Date Line, indicating less emission of heat in the ENSO year winter than normal years. In the western equatorial area strong negative anomalies appear, while anomalies are positive in the central equatorial area.

Although the PNA pattern is likely to appear in an ENSO year winter, it is not always the case. For instance, in the winter of 1972/1973 during an ENSO event, no PNA pattern was generated. A correct understanding may be that ENSO prepares conditions favorable for the PNA pattern to occur. How the mid and high latitude atmosphere independently behaves, or how the ocean–atmosphere interaction in mid and high latitudes prevents or assists forcing from the tropical region is not well clarified yet.

b) PJ pattern and summertime climate

After the rainy season Japan is covered with a subtropical high pressure and the southeast monsoon wind starts to dominate. Climate in summer is largely controlled by the size of this high pressure. The PJ (Pacific–Japan) pattern is known as a teleconnection that intensifies or weakens the western part of the subtropical high (Nitta, 1987). A schematic diagram of the pattern is shown in Fig. 4-15. A wave train is shown along the western coast of the Pacific from the warm pool area in the tropics. When a warm water is massed in the western Pacific with high sea surface temperature (La Niña), high pressure anomalies cover all of Japan. This corresponds to a thoroughly developed subtropical high

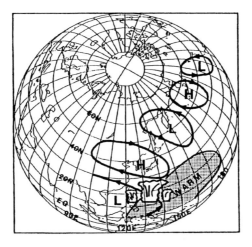

Fig. 4-15. Schematic picture showing the relationships among SST anomalies, convective activities and atmospheric Rossby-wave trains (Pacific–Japan teleconnection pattern). After Nitta (1987).

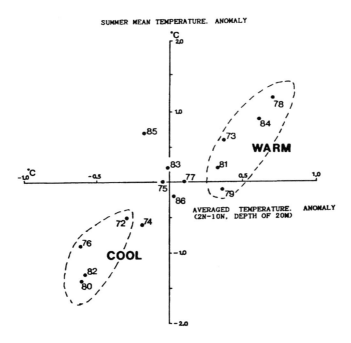

Fig. 4-16. Scatter diagram of the tropical ocean temperature and the summer mean temperature in western Japan. The warm (cold) year group consists of years in which the ocean temperature anomalies are equal to or greater (smaller) than 0.52 (–0.52) standard deviation. After Kurihara (1989).

pressure extending to the west. The pressure pattern brings about extremely hot summer in Japan. On the other hand, when seawater temperatures are high in the western tropics (El Niño), pressure anomalies are reversed and Japan is covered with low-pressure anomalies. In such cases, subtropical high pressure is less developed and Japanese summer tends to be cool.

Figure 4-16 exhibits the quantitative relationship between the seawater temperatures in the surface layer of the western tropics and the temperatures of western Japan in summer (Kurihara, 1989). It is seen that the two events are correlated. 1972, 1976 and 1982 were years when ENSO events occurred, corresponding to cool summer. In La Niña years (low temperature anomalies in the western Pacific), the western part of Japan had very hot summers. However, it should be repeated that saying "cool summer for El Niño and hot summer for La Niña" is too simplistic. This is because, as mentioned in the previous section, the mid latitude atmosphere also has its own dynamics, independent from ENSO events. Nevertheless, ENSO events are still an important factor for climate control in Japan.

4.5 LARGE-SCALE OCEANIC VARIATIONS AND ATMOSPHERIC CIRCULATION

In Section 4.2 we discussed the importance of the role of the ocean in climate change, although in a rather conceptual and abstract way. This section will deal with the recently revealed facts about large-scale variations of the ocean and atmosphere associated with short-term climate change, using the observed data. However, this field is still in the process of rapid development and we are also going to present some subjects whose assessments have not been properly established yet.

Not many attempts have been made to detect long-term variations in the ocean on the large scale. Sea level observations (by tide gauge), sea surface temperatures, data from a station (e.g., Station "S"; Station "P") and an observation line (e.g., 137°E by the Japan Meteorological Agency), and surface layer temperatures (by XBTs) are the main long-term observations. As these data are accumulated and stored, long-term variations are being clarified, even if fragmentarily. In the succeeding sections we firstly discuss long-term variations detected in sea surface temperatures, which are the most abundant of all oceanographic data.

4.5.1 Variations of sea surface temperatures in the North Pacific

Analyses of sea surface temperatures (SST) have been made since the older days. As data sets were more systematically accumulated, more oceanic and global scale analyses have been carried out. New methods of analysis for detecting a signal have been introduced into oceanography, too.

Numerous researchers now apply the EOF (Empirical Orthogonal Function) analysis to various fields to obtain the variation pattern and its temporal change. Davis (1976) and Weare *et al.* (1976) were the first to apply the method to SSTs

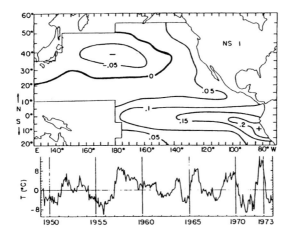

Fig. 4-17. Distribution map of the leading EOF mode of SST anomalies. This mode can account for 23.1% of the total variance. After Weare *et al.* (1976).

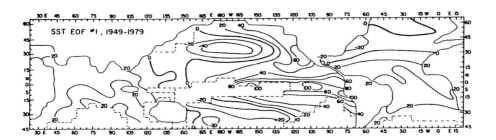

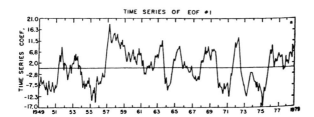

Fig. 4-18. Distribution map of the leading EOF mode of global SST anomalies for 1949–1979. This mode can account for 7.67% of the total variance. After Hsiung and Newell (1983).

in the North Pacific. Weare *et al.* performed an EOF analysis on monthly sea surface temperatures between 55°N and 20°S during the period 1949–1973. The pattern of the EOF first mode together with its corresponding time coefficients is shown in Fig. 4-17. There is an area of large amplitude in the upwelling region near the equator off South America, whereas a western part in the middle latitude has the opposite sign: they are anti-correlated. The years of large positive amplitude of the time coefficients correspond to the occurrences of the ENSO event. In this sense the ENSO was identified. Hsiung and Newell (1983) extended the analysis to the entire earth. The leading mode of their results is shown in Fig. 4-18. The pattern in the Pacific and variation of time coefficient is very similar to those of Weare *et al.* Therefore, this leading mode can be regarded as the ENSO mode.

Kawamura (1984) and Iwasaka *et al.* (1987) also made the EOF analysis of SST and studied its relationship with the resultant variations of atmospheric circulation fields. They found that the variation of the sea surface temperatures extracted as the first EOF mode is related to the activities of the PNA teleconnection pattern.

It is seen from Fig. 4-18 that in addition to the ENSO events, a variation of a longer period seems to be present. Tanimoto *et al.* (1993) filtered the data and made the EOF analysis for separated frequency bands. A variation with four-year period in average is observed, corresponding to the intervals of the ENSO events (2–5 years). Figure 4-19 shows the distribution of the first EOF of SST for this time scale in the North Pacific (upper panel) and its time coefficient (lower panel). It is to be noted that the center of action lies at (35°N, 150°W), a little east of the corresponding center in Figs. 4-17 or 4-18, and the western Pacific is covered by anomalies with the opposite polarity. Most of the significant positive values of the time coefficient correspond to the ENSO cold events (La Niña; 1955/56, 1966/67, 1970/71, 1980/81, and 1983/84), and most of the negative values to the ENSO warm events (El Niño; 1957/58, 1965/66, 1968/69, 1972/73, and 1982/83). This result confirms that the North Pacific SST anomalies with this time scale are strongly influenced directly or indirectly by tropical atmosphere-

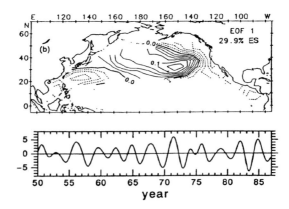

Fig. 4-19. Distribution map of the leading EOF mode of SST anomalies with ENSO time scale (2 year to 5 year) in the Pacific (upper panel) and its time coefficient (lower). This mode can account for 29.9%. After Tanimoto *et al.* (1993).

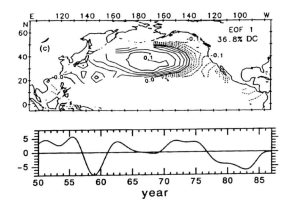

Fig. 4-20. As in Fig. 4-19 except for that of SST anomalies with decadal time scale (longer than five years) in the Pacific (upper panel) and its time coefficient (lower). This mode can account for 36.8%. After Tanimoto *et al.* (1993).

ocean interactions, although they might be somewhat modulated by interactions within the extratropics.

Figure 4-20 shows the result of the EOF analysis, using only the variations that have periods of five years or longer. The distribution in this figure rather resembles that of mid latitudes in Fig. 4-17 or Fig. 4-19. The center of action is near the international date line at 35°N. The time coefficients do not show regular periodic variations. We can recognize a quasi-steady anomaly period immediately followed by another persistent anomaly period with the opposite polarity, with rather abrupt transitions occurring around 1956, 1960 and 1976. Figure 4-21 shows composite maps based on these time coefficients, after the periods for the anomalies are divided into three categories: the periods of large positive values (L+), of large negative values (L–) and the remaining period (L0). For the periods of the L+ and L–, the extratropical SST anomalies are characterized by the pattern associated with the first EOF, while the anomalies with opposite polarity dominate over the tropics at least for the L+ period. The result indicates a decadal fluctuation in the equator to the pole gradient of SST over the Pacific Ocean. The L0 period pattern is characterized by noisy, almost normal SST anomalies in the extratropics and a large positive anomaly in the southeastern part of the domain.

4.5.2 Transition in the mid 1970s

Let us note the transition that occurred in the middle of the 1970s. It was pointed out in the late 1980s that the wintertime atmospheric general circulation field over the north Pacific had shifted to a different state than before. The results of the studies all conclude that, from the mid 1970s through the 1980s, El Niño occurred more frequently than before and no La Niña existed, so that the Aleutian Low was developed by PNA teleconnection and kept migrating to the east.

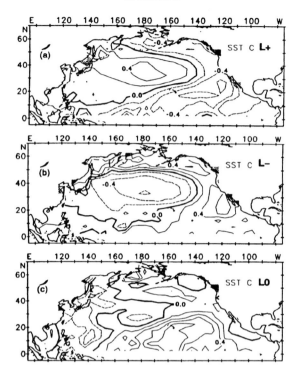

Fig. 4-21. Composite maps for unfiltered SST anomalies for three categorized periods based on the time coefficient shown in Fig. 4-20: (a) L+ (period with large positive time coefficient), (b) L– (large negative), and (c) L0 (the other). Contour interval is 0.2°C. Positive (negative) values are represented by solid (dashed) lines. After Tanimoto *et al.* (1993).

Kashiwabara (1987) was the first to show that the wintertime 500 hPa height significantly lowered over the north Pacific around 1977. The center of the lowering was near (40°N, 160°W); the difference from the mean height of 30 years from 1951 to 1980 was as large as 80 m. On the other hand the height over Alaska went up by more than 40 m. Kashiwabara also showed time series of Southern Oscillation, pointing out its relationship with convective activities in the tropics and inferred that the variations in the mid- and high latitudes reversibly affect the tropical region.

Nitta and Yamada (1989) studied decadal to interdecadal variations, using SST and outgoing long wave radiation data for the whole globe. As a result they found that temperatures were higher after the late 1970s in the tropical Pacific, and associated convection became more active. Consequently, the PNA pattern distinctly appeared in the atmosphere, causing a considerable lowering of the 500 hPa height over the north Pacific. Contrary to Kashiwabara, they attributed the main player of the variations to the tropical SSTs.

Trenberth (1990), focusing on the warming in the western coast of North America, particularly near Alaska, reached the same scenario as Nitta and Yamada. For 10 years from 1977 to 1988 strong El Niño took place three times and no La Niña. This, through the PNA pattern, suppressed the 500 hPa height over the north Pacific while elevating it over Alaska. Hence, cold air is advected to the mid and high latitudes in the Pacific, lowering SSTs, while humid and warm air moves to Alaska raising the air temperature there.

Such transition in the mid 1970s is also found in oceanic variations. Sverdrup transports using wind stress fields indicate that the volume transports of both subtropical and subarctic circulations would increase from the mid 1970s (Hanawa, 1995).

4.5.3 Long-term variations in subsurface temperature field

What kind of vertical structure does the signal in an SST field shown in the previous subsection have? Recently, an analysis of this matter has been started using historically accumulated data. Figure 4-22 shows a time–longitude diagram of temperature vertically averaged from the surface to 400 m (right panel) and SST (left), which are also zonal means from 35°N to 45°N (Watanabe and Mizuno, 1994). The former variable can be regarded as heat storage. We can observe an abrupt change in the mid 1970s in both heat storage and SST fields.

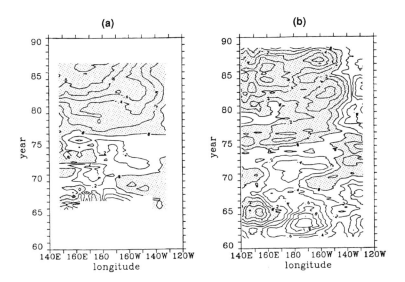

Fig. 4-22. (a) Time longitude matrix of 0–400 m vertically averaged temperature anomalies along the 35–45°N latitude band. Time series at each grid point are a 3-year running average. Contour interval is 0.2°C and negative values are shaded. (b) Same as (a) except for SST. SST data set was provided by the Japan Meteorological Agency. After Watanabe and Mizuno (1994).

That is, we can judge that the signal also appears in the subsurface temperature field with certain strength. Watanabe and Mizuno also made difference maps of both fields for two ten-year periods of 1966–1975 and 1976–1985 (latter minus former, not shown here). As expected, the pattern of the SST difference map was very similar to the leading EOF mode of SST anomalies shown in Fig. 4-20. On the other hand, the heat storage difference map showed the negative signed region in the mid-latitudes was shifted more southward compared with SST fields, and positive (negative) heat storage appeared in the eastern (western) equatorial region. This means in the latter period, warm water was relatively shifted eastward.

Figure 4-23 shows the distribution of 2-year lag correlation coefficients between 400 m temperature field and 10 m temperature at the reference station of 32.5°N and 172.5°W. Based on the temperature cross section (not shown here) and Fig. 4-23, Watanabe and Mizuno inferred that the anomalies first appeared in the 30–40°N zonal belt, subducted to deeper layer as time progressed, and circulated southward in the upper thermocline. This might be the manifestation of a subduction process in the ventilated thermocline theory (Luyten *et al.*, 1983). Although Talley (1985) regarded a shallow salinity minimum observed in the northeastern part of the North Pacific subtropical gyre as the signal of subduction, we can conclude that Watanabe and Mizuno could extract the propagation signal of the subducted water.

Cox and Bryan (1984) showed, using the ocean general circulation model, that the time scale of the subduction process is several years to several decades for an ocean the size of the North Atlantic. The periodicity of decadal-to-interdecadal time scale variations mentioned in the previous section might be determined by this subduction process. That is, the following scenario (working

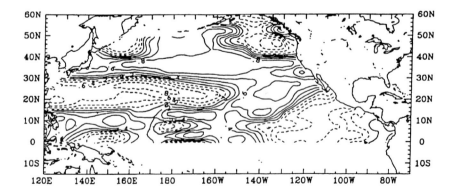

Fig. 4-23. One point correlation map for 400 m temperature at a two-year lag with the reference time series of 10 m temperature at 32.5°N, 172.5°W. Contour interval is 0.2 and negative contours are dashed. After Watanabe and Mizuno (1994).

hypothesis) can be proposed. First, let us consider the initial state that equatorial SST anomalies are positive. Then, due to positive SST anomalies in the equatorial region, a PNA teleconnection pattern would immediately be excited in the atmosphere. Then, this atmospheric circulation would force SSTs in the mid-latitude ocean to cooler (negative anomalies). This would also bring the formation of a cold water mass. This cold water mass slowly subducts to a deeper layer and moves to the equatorial region in the upper thermocline with the decadal-to-interdecadal time scale. When this cold water mass reaches the equatorial region, it influences the upper ocean stratification and changes the SST anomaly field from positive to negative. Then, in this case, an anti-PNA teleconnection pattern would be excited and make positive SST anomalies in the mid-latitude ocean. Due to this atmospheric forcing, SSTs in the mid-latitude ocean take a positive anomaly, and a warm water mass is formed. This warm water also subducts slowly and moves to the equatorial region and changes the SST anomaly field from negative to positive. This state is just the first one: the oscillation is closed. Since the subduction process plays a key role in this scenario and also determines the time scale of this oscillation, we may call this oscillation "Subduction Oscillation".

Whether or not this working hypothesis mentioned above actually plays in the real atmosphere–ocean system should be examined in the future. In order to examine this hypothesis, we may need a comprehensive project including observational process study on subduction. An analysis of temporal variability in oceanic subsurface structure has just started. Although the data is not necessarily adequate for this purpose, the author believes we can extract much useful information by careful examination of the historically accumulated data.

4.5.4 "Climate jump"

In relation to the interdecadal variations discussed above, a concept of a "climate jump" or "regime shift" has been introduced and applied. Its simple description will be given in this section.

As pointed out by Lorenz (1963), a nonlinear system of deterministic equations can produce "chaos" with no external perturbations. His model originally represented the modulation of intensity of the Benard convention, expanding the fields of motion and temperature into horizontal and vertical wave numbers, and cutting them by a small number of dimensions. The derived system of governing equations is nonlinear ordinary simultaneous equations in three dimensions. He found, after applying various initial conditions to the equations, that the solution within a certain parameter range behaves like chaos moving non-periodically around two stationary solutions. He conjectured that variations of the atmosphere comprise such a nonlinear system and its long-term forecasting might not be possible.

Based on this work, Lorenz (1968) attempted to classify the climate systems. One is the transitive system, which asymptotically approaches an equilibrium solution, independently of its initial state. Another is the intransitive system, which allows two or more stable solutions, but one solution is determined by the initial condition. The third is the almost-intransitive system, a special type of the first one, which transits from one solution to another with one stochastic state during a certain period being different from another stochastic state during another period. He proposes that the present climate system is an almost-intransitive one.

Charney and De Vore (1979) reproduced an example of the almost-intransitive system, by studying an east-west belt of flow, corresponding to westerlies. Their investigation is on the multi-equilibrium nature of zonal flow solutions and their transitions, to understand the blocking phenomenon. Adopting a barotropic quasi-geostrophic model, they showed that there exist two stable solutions. One is the "high index" state where a zonal flow dominates, and the other the "low index" state where a meander is predominant. In other words, in the time-evolution system, the state fluctuates around a certain stable solution for a certain finite period, and after an abrupt transition it fluctuates again around another stable solution. This is an example of an almost-intransitive system. If the climate system is such an example, the abrupt transition can be called a "climate jump". However, since this theory uses a model with a very limited degree of freedom, how well it represents the real atmosphere, whose degree of freedom is almost infinite, remains to be questioned.

Following the theoretical works, Yamamoto *et al.* (1986) attempted to determine if a climate jump is recognized in observational data. The climate jump here is defined for "the year when the difference of mean values for the previous ten or more years and for the following ten or more years is larger than the statistically significant difference (the ratio of the signal to the noise is bigger than one): the difference between the two mean values divided by the sum of the two standard deviations is larger than one". They reported that climate jump took place around the 1950s in several elements of the meteorological data from 1900 to the 1970s in Japan.

4.6 IMPORTANCE OF SEA SURFACE FLUX ANALYSIS

In spite of the title of this chapter, we have not been able to describe the real contents of the "interactions" in a strict sense of the word. In fact, although some mechanism of variations in the tropics where the interactions take place locally (or vertically), such as ENSO, has been clarified, the interactions in the mid and high latitudes are not localized so that they cannot be well understood from data analysis only. Most studies simply point out that the ocean and the atmosphere vary coherently and nothing more. Generally speaking, meteorologists tended to say "the atmosphere changed in response to the changed SST, which functions as the lower boundary condition", whereas oceanographers were likely to advocate "the ocean changed in response to the changed circulation in the atmosphere". Nowadays, of course, many researchers are dissatisfied with such uni-directional reasoning, and are making their best efforts to elucidate the "true interactions".

Studies on SSTs have been done extensively for various reasons. The results are fruitful. However, the concept here is "atmosphere–SST–ocean", namely, SST is considered to play a connecting role between the atmosphere and the ocean. The present author believes that more researches from a viewpoint of "atmosphere-sea surface flux–ocean" should be made from now on.

Sea surface fluxes are obtained by combining observed physical quantities as discussed in Section 4.3. There are not so much observed data as SSTs, and flux data should be treated with care. SSTs are associated with the temporal rate of the change of heat stored in the mixed layer. On the other hand, the thermal flux across the sea surface varies much more than the sea surface temperature does, because it is a factor that causes the rate of heat change. We could say this the other way round, that is, the sea surface temperature can be easily observed because it is proportional to an integrated quantity of heat flux. However, the present author thinks that a next breakthrough lies in the analysis and interpretation of sea surface fluxes.

The research in this field seeks understanding of the mechanism of climate change as a means for distinguishing artificial factors from natural ones in the global warming problem. More and more researchers are involved in this field, which is developing very rapidly. The contents described in this chapter may become old tomorrow and more new findings may have to be added then. The author expects such necessity of revision in the near future.

REFERENCES

Charney, J. G. and J. G. De Vore (1979): *J. Atmos. Sci.*, **36**, 1205–1216.
Cox, M. D. and K. Bryan (1984): *J. Phys. Oceanogr.*, **14**, 674–687.
Davis, R. E. (1976): *J. Phys. Oceanogr.*, **6**, 249–266.
Hanawa, K., *et al.* (1988): *J. Meteor. Soc. Japan*, **66**, 445–456.
Hanawa, K. (1993): *Kisho Kenkyu Note*, **180**, 31–94 (in Japanese).
Hanawa, K. (1995): *Hokkaido Nat. Fish. Res. Inst.*, **59**, 103–120.
Hellerman, S. and M. Rosenstein (1983): *J. Phys. Oceanogr.*, **13**, 1093–1104.
Horel, J. D. and J. M. Wallace (1981): *Mon. Wea. Rev.*, **109**, 813–829.
Hsiung, J. and R. E. Newell (1983): *J. Phys. Oceanogr.*, **13**, 1957–1967.
Ishii, M., *et al.* (1994): *Geophys. Mag.*, **45**, 19–54.
Iwasaka, N., *et al.* (1987): *J. Meteor. Soc. Japan*, **65**, 103–113.
Kashiwabara, T. (1987): *Tenki*, **34**, 777–781 (in Japanese).
Kawamura, R. (1984): *J. Meteor. Soc. Japan*, **62**, 910–916.
Kurihara, K. (1989): *Geophys. Mag.*, **43**, 45–104.
Kutzbach, J. E. (1976): *Quater. Res.*, **6**, 471–480.
Lorenz, N. N. (1963): *J. Atmos. Soc.*, **20**, 130–141.
Lorenz, N. N. (1968): *Meteor. Monogr.*, **8-30**, 255–267.
Luyten, J. R., *et al.* (1983): *J. Phys. Oceanogr.*, **13**, 297–309.
Manabe, S., *et al.* (1991): *J. Climate*, **4**, 785–818.
Nitta, T. (1987): *J. Meteor. Soc. Japan*, **65**, 373–390.
Nitta, T. and S. Yamada (1989): *J. Meteor. Soc. Japan*, **67**, 375–383.
Oberhuber, J. M. (1988): *Max-Planck Insti. Meteor. Rep.*, **15**.
Shea, D. J. (1986): NCAR, Rep.
Talley, L. D. (1985): *J. Phys. Oceanogr.*, **15**, 633–649.
Tanimoto, Y., *et al.* (1993): *J. Climate*, **6**, 1153–1160.

Trenberth, K. E. (1990): *Bull. Amer. Meteor. Soc.*, **71**, 988–993.

Tsunogai, S. (1981): *Chikyu Kagaku*, **15**, 255–267 (in Japanese).

Walker, G. T. and E. W. Bliss (1932): *Mem. Roy. Meteor. Soc.*, **4**, 53–84.

Wallace, J. M. and D. S. Gutzler (1981): *Mon. Wea. Rev.*, **109**, 784–812.

Watanabe, T. and K. Mizuno (1994): *Int. WOCE Newslett.*, **15**, 10–14.

Weare, B. C., *et al.* (1976): *J. Phys. Oceanogr.*, **6**, 671–678.

Yamamoto, R., *et al.* (1986): *J. Meteor. Soc. Japan*, **64**, 273–281.

Chapter 5

Fundamentals of Large-Scale Interaction

Toshio YAMAGATA and Yoshinobu WAKATA

Ocean–Atmosphere Interactions, Ed. Y. Toba, pp. 143–193.
© by TERRAPUB / Kluwer, 2003.

Fundamentals of Large-Scale Interaction

Toshio YAMAGATA and Yoshinobu WAKATA

The fountains mingle with the river,
And the rivers with the ocean,
The winds of heaven mix for ever, ...

—*Percy Bysshe Shelley*

5.1 EL NIÑO AND SOUTHERN OSCILLATION

Why do we blow on tea to cool it? A thought on such a casual action we make in daily life may lead to a clue for understanding ocean–atmosphere interactions.

If we leave hot tea without doing anything, its temperature will not drop readily. Radiation, thermal conduction by molecules in the air, or feeble convection occurring over the hot tea will not be very effective to cool it. That is why we drive away damp air near the surface to activate more evaporation and take away its latent heat. We can also expect to generate turbulence in the air to transport more sensible heat. In addition, we cannot overlook the fact that blowing will make on the surface small irregularities susceptible to surface tension, will create a flow, and will transfer its momentum from the air to the tea more efficiently, thus causing a forced convection, overturning hot water below and cooled water in the suface layer.

Similarly, winds blowing over the sea suface deprive the ocean of its heat through the sea surface boundary processes with evaporation, while they exchange momentum quite effectively with the ocean by generating wind waves, whose wavelengths and amplitudes are larger than those of capillary waves. When we want to cool tea, we provide wind intentionally from outside, but nature is self-sufficient. Seawater vapor that was evaporated from the sea suface eventually condenses and releases latent heat into the air. The heated air rises and the compensating converging (or diverging) movement of the air will form a circulation in the atmosphere. The wind associated with the circulation generates currents in the ocean surface layer through sea suface boundary processes. Where the surface currents converge, warm water accumulates and suppresses cold water from deeper layers, thus becoming even warmer and producing more evaporation. In this way, large-scale dynamic and thermodynamic processes of the atmosphere and the ocean influence one another without ceasing. It is surprising that before the early 1980s, most oceanographers were studying the response of the ocean under given conditions

of wind and heat flux from the atmosphere, whereas meteorologists were investigating atmospheric circulation under given sea surface temperatures.

In tropics where temperature is high and humidity is large, ocean–
atmosphere interactions are especially active. The most outstanding example is
the El Niño phenomenon in the tropical ocean. El Niño (meaning the Child) was
originally the word used by local seamen to indicate the warm current coming
from the equator off Peru when the cold Peru (Humbolt) Current weakens around
Christmas time. However, once in several years (2–9 years) this warm water
spreads over the entire eastern tropics and lasts for a long period of time (about
one year). Such large-scale El Niño affects not only the ocean but also the
atmosphere. Drought, torrential rain, warm winter, and abnormal occurrence of
typhoon, hurricanes, or cyclones, etc. have become objects of social attention,
too. On the other hand, meteorologists knew from a long time ago about a
phenomenon called Southern Oscillation. This, discovered fortuitously by Sir
Walker who was trying to forecast irregular monsoons in India, is seemingly a
strange phenomenon where surface pressure at Darwin in Australia and surface
pressure at Tahiti in the Pacific co-oscillate with a period of a few years. In
particular, the difference between the surface pressure anomalies at Darwin and
at Tahiti is called the South Oscillation Index, which is known to be anti-
correlated with the time series of seawater temperatures off Peru, which
represents El Niño (Fig. 5-1). It was not until the 1980s that Southern
Oscillation of the tropical atmosphere and El Niño of the tropical ocean were
seen to be two aspects of a single phenomenon. This large-scale connection
between the atmosphere and the ocean in the tropics is now called ENSO (El Niño
and Southern Oscillation).

In this chapter we will introduce a new field of earth science which merges
physical oceanography and meteorology, with its main emphasis on the ENSO

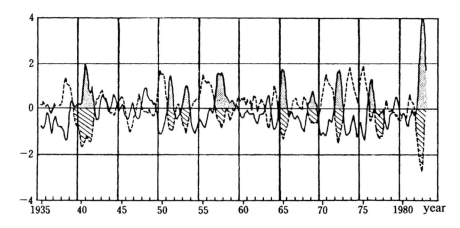

Fig. 5-1. Southern Oscillation Index (pressure difference between Tahiti and Darwin)
(broken line), and sea temperature near Puerto Chicama in Peru (solid line).

phenomenon. It is known theoretically and observationally that a variety of equatorial waves exist in the tropical ocean and atmosphere. The basis of this chapter is a concept of waves both in the ocean and atmosphere. An instability theory of the coupled ocean–atmosphere system was introduced to explain warming of seawater temperature accompanying occurrence of El Niño and increasing westerly component progressing eastward which is associated with the ENSO event (Philander *et al.*, 1983; Yamagata, 1985; Hirst, 1986). When we make an attempt to formulate the ocean and the atmosphere as a single coupled system, equatorial waves in the ocean and atmosphere are coupled and a new type of wave which amplifies with time appears. This unstable mode, which is produced by a positive feedback mechanism characteristic of the coupled ocean–atmosphere system in tropics, is the very origin of El Niño.

As is evident from Fig. 5-1, El Niño does not continue to retain its original form after its occurrence. The western part of the Pacific eventually becomes warm again, and the eastern Pacific becomes cold (sometimes called La Niña), then another El Niño phase starts. The interval of the occurrence is not regular (2–9 years), but the temperature in the east Pacific goes up and down every several years. Hence, we should elucidate the mechanism of transition from the warm state to the cold state or vice versa, in addition to the amplification mechanism. It is essential to clarify this transition mechanism for forecasting the El Niño. Two theories have been proposed so far. One is that equatorial waves generated by the preceding El Niño or La Niña play a main role. The other is that in relation to the Asian monsoon, the convection over the maritime continent (Indonesian Sea, Borneo, Sumatra, etc.) is an important factor. Although some attempts are being made to forecast the ENSO phenomenon using coupled ocean–atmosphere models, it has not yet been fully understood and still is an interesting subject of research.

In Section 5.2, we are going to derive fundamental equations that govern the equatorial ocean and will account for the dynamics in the tropics focusing on equatorial waves. Oceanic circulation in the tropics will be explained by studying forced solutions of the equatorial waves. In Section 5.3, a simple model of tropical atmosphere will be formulated. In Section 5.4, a theory of coupled instability, in which anomalies of seawater temperatures and winds increase due to their mutual interactions, will be presented as a cause of ENSO. In Section 5.5, a simple ocean–atmosphere coupled model (delayed oscillator model) will be introduced to study the mechanism to reproduce a life cycle of ENSO from its begiming to end. In Section 5.6, the predictability of ENSO and future problems will be discussed.

5.2 FUNDAMENTALS OF OCEAN DYNAMICS IN THE TROPICS

5.2.1 Fundamental equations of ocean dynarnics

In this section, equations of motions for ocean dynamics in the tropics will be derived. For simplicity, we will ignore advective terms and linearize the equations. Under Boussinesq approximation[1] and hydrostatic approximation[2] the equations on a local Cartesian coordinate system will be

$$u_t - fv + p_x = v_H \nabla^2 u + \left(v_V u_z \right)_z + F^x \tag{5-1}$$

$$v_t + fu + p_y = v_H \nabla^2 v + \left(v_V v_z \right)_z + F^y \tag{5-2}$$

$$p_z = -\rho g \tag{5-3}$$

where subscripts x, y, z and t indicate partial derivatives. u and v are east-west and south-north velocities, p and ρ are the density anomalies divided by the mean density ρ_0 of sea water. v_H and v_V are horizontal and vertical eddy viscosity coefficients, respectively. F^x and F^y indicate external forces such as wind stress. f is the Coriolis parameter; in the equator $f = \beta y$.

The equation of continuity under the same assumptions will be

$$u_x + v_y + w_z = 0 \tag{5-4}$$

where w is the vertical velocity. The equation that governs the density will be

$$\rho_t - \frac{N^2}{g} w = \left(\kappa_V \rho_z \right)_z \tag{5-5}$$

where κ_V is the vertical eddy diffusivity coefficient of density. N^2 is equivalent to $-g/\rho_0 d\rho/dz$ and N is called Brunt–Väisälä frequency. Assuming the fluid to be incompressible, the density ρ can be expressed as a function of temperature T: $\rho = \rho_0(1 - \alpha T)$, where α is the thermal expansion coefficient.

The boundary condition at the flat ocean bottom ($z = -D$) will be

$$w = 0 \tag{5-6}$$

If we impose free boundary condition at $z = 0$, ignoring the vertical displacement of the free surface, we obtain[3]

---

[1] The density of the fluid under consideration is assumed to be constant, except the buoyancy term. Under this approximation, the fluid is assumed to be incompressible and the vertical scale of the phenomenon is assumed to be smaller than the scale height of the basic stratification.

[2] In the vertical direction, the gravity force and the pressure gradient force are assumed to be in balance, neglecting vertical acceleration.

$$w - \frac{1}{g} p_t \cong 0 \qquad (5\text{-}7)$$

For the sake of simplicity, density variations at the surface and at the bottom will be ignored.

It is difficult to solve these equations analytically, but if the system is non-dissipative (Lighthill, 1966), or if diffusion is only in the vertical direction and the relevant coefficients can be expressed by the Brunt–Väisälä frequency as $\nu_H = A/N^2$ and $\kappa_V = B/N^2$ (A and B are constants) and if the right-hand side of Eq. (5-5) is replaced by $(\kappa_V \rho)_{zz}$ for the sake of mathematical expediency, then vertical and horizontal variables can be separated (McCreary, 1981). In this case, the solution can be decomposed into an infinite number of eigenfunctions, and their separation constants or eigenvalues can be determined from boundary conditions. Now we expand each variable as follows:

$$u = \Sigma u_m(x, y, t) R_m(z), \qquad (5\text{-}8\text{a})$$

$$v = \Sigma v_m(x, y, t) R_m(z), \qquad (5\text{-}8\text{b})$$

$$p = \Sigma p_m(x, y, t) R_m(z), \qquad (5\text{-}8\text{c})$$

$$w = \Sigma w_m(x, y, t) S_m(z), \qquad (5\text{-}8\text{d})$$

$$\rho = \Sigma \rho_m(x, y, t) \frac{dR_m(z)}{dz}. \qquad (5\text{-}8\text{e})$$

The external force can also be expanded as

$$F_m^{(x,y)} = \frac{\int_{-D}^{0} F^{(x,y)}(x, y, z, t) R_m dz}{\int_{-D}^{0} R_m^2 dz}. \qquad (5\text{-}9)$$

Now using the fundamental equation system, we can derive the equations

---

[3]The free boundary surface condition can be obtained as follows. Since the pressure at the sea surface is equal to the atmospheric pressure (which is assumed to be constant), the material derivative in the vertical direction of the pressure at the surface becomes null when horizontal motion is small. That is: $\partial p/\partial t + w(\partial p/\partial z) = 0$. Linearizing this equation by the use of the hydrostatic relationship in the mean field $\partial \bar{p} / \partial z = -\bar{\rho} g \sim \rho_0 g$ (here $\bar{p}$ and $\bar{\rho}$ are pressure and density not divided by ρ_0 respectively) will lead to Eq. (5-7).

that govern the m-th component of the horizontal structure functions:

$$u_{mt} - f v_m + p_{mx} = -\frac{A u_m}{C_m^2} + F_m^x , \tag{5-10a}$$

$$v_{mt} + f u_m + p_{my} = -\frac{A v_m}{C_m^2} + F_m^y , \tag{5-10b}$$

$$p_{mt} + C_m^2 \left(u_{mx} + v_{my} \right) = -\frac{B p_m}{C_m^2} , \tag{5-10c}$$

where C_m is a separation constant. C_m is arranged in descending order as C_0, C_1, C_2, ..., C_0 being the largest. If we replace p with $g\eta$, and C_m^2 with gH_m, then the above equations will be linearized shallow-water equations for the ocean of depth H_m and the vertical displacement η of the sea surface. Since the phase velocity of the long gravity wave component in this system of shallow-water equations is the square root of gH_m, the phase velocity of the m-th gravity wave mode is equal to the eigenvalue C_m.

The vertical structure function is determined by

$$C_m^2 \frac{d^2 S_m}{dz^2} + N^2 S_m = 0 \tag{5-11}$$

and S_m and R_m are related by

$$R_m = -C_m^2 \frac{dS_m}{dz} . \tag{5-12}$$

The equation of continuity for each vertical mode becomes

$$C_m^2 \left(u_{mx} + v_{my} \right) - w_m = 0 . \tag{5-13}$$

The hydrostatic relation is

$$p_m = -\rho_m g . \tag{5-14}$$

The density is governed by

$$\rho_{mt} - \frac{1}{g} w_m = -\frac{B}{C_m^2} \rho_m . \tag{5-15}$$

The constant of separation can be determined using the upper and lower boundary conditions

$$w = S_m = 0 \qquad\qquad z = -D, \qquad (5\text{-}16)$$

$$w - \frac{1}{g}p_t \cong S_m - \frac{C_m^{\ 2}}{g}S_{mz} = 0 \qquad z = 0. \qquad (5\text{-}17)$$

Let us consider the case where C_m is very large. When $N^2(z)$ is assumed to be constant, the solution will be of the form $S_0 \sim \hat{A}\sin(Nz/C_0) + \hat{B}\cos(Nz/C_0)$, where C_0 should satisfy the boundary condition $C_0 N/g = \tan(ND/C_0)$. If $C_0 \gg ND$, then $\tan(ND/C_0) \sim (ND/C_0)$ so that the eigennvalue for this mode will be determined as $C_0 = (gD)^{1/2}$. C_0 is equal to the phase velocity of the external gravity wave in shallow water of depth D and it is much larger than ND. The variations of the density and temperature associated with this mode are small. The displacement of the sea level (or the pressure variation at $z = 0$) makes the pressure variation uniform in the vertical direction and therefore horizontal velocities are almost uniform in the vertical direction. This mode is known as the barotropic mode. Higher modes ($m \geq 1$) are called baroclinic modes, for their pressure variations are caused mainly by density variations, so that isopycnic surfaces and isobaric surfaces intersect each other. Then C_m is small and the upper boundary condition can be approximated as $S = 0$. If we integrate R_m vertically and use the boundary condition together with Eq. (5-12), we obtain

$$\int_{-D}^{0} R_m dz = -C_m^{\ 2}[S(0) - S(-D)] = 0. \qquad (5\text{-}18)$$

That is, the vertical integrated horizontal velocity and pressure variation of baroclinic modes are zero.

Figure 5-2 is representative of the observed temperature and stratification of the tropical oceans, which varies considerably with time and position, but which is always such that the Brunt–Väisälä frequency has a maximum in the thermocline at a depth of about 100 m. Figure 5-3 shows the structure of the first few vertical modes corresponding to the vertical temperature distribution. As the vertical mode number m increases, the number of nodes of the mode increases so that m can be regarded as a vertical wave number. Table 5-1 lists the first few eigenvalues C_m, the corresponding equivalent depths $H_m = (C_m^{\ 2}/g)$, the representative spatial scales (the equatorial radii of deformation) $L_m = (C_m/\beta)^{1/2}$ and the representative time scales (the equatorial inertial time) $T_m = (C_m\beta)^{-1/2}$. It is shown that the barotropic mode ($m = 0$) has an equivalent depth equal to the actual ocean and the baroclinic modes have equivalent depths much smaller than the depth of the ocean. As m increases, both the phase speed and the horizontal scale decreases. As is seen from Fig. 5-3, the amplitudes of baroclinic

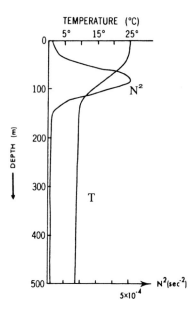

Fig. 5-2. Vertical profile of temperature and Brunt–Väisälä frequency representing the tropical ocean (Philander, 1990). The temperature under 500 m is reduced with a constant rate up to the ocean floor of 4000 m.

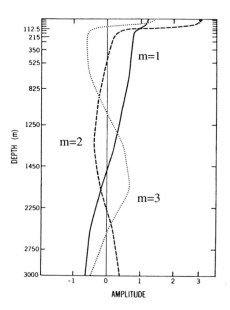

Fig. 5-3. Vertical structure of the baroclinic modes obtained from the vertical profile of the temperature in Fig. 5-2 (Philander, 1990).

Table 5-1. Equivalent depth, eigenvalue, representative spatial scale and temporal scale for each vertical mode in an ocean with a vertical profile of the temperature in Fig. 5-2.

m	H_m (cm)	C_m (cm s^{-1})	L_m (km)	T_m (days)
0	400,000	20,000	3,000	0.2
1	60	240	325	1.5
2	20	140	247	2.0
3	8	88	197	2.6
4	4	63	165	3.1
5	2	44	131	3.6

modes vary mainly in the surface layer. The variations of pressure and velocity components will follow the vertical structure shown in the figure, as can be deduced from Eqs. (5-8a), (5-8b), (5-8c) and (5-8d). However, the density (or temperature) varies a little differently, because it is determined by the vertical gradient of the amplitude function in Fig. 5-3. The variation becomes large near the picnocline (thermocline).

5.2.2 Equatorial waves

The system of Eq. (5-10) allows us to obtain the characteristics of equatorial waves. We will pay attention to one vertical mode, omitting the vertical mode number m unless necessary. If the equations are nondimensionalized by the temporal scale $T = (C\beta)^{-1/2}$ and the spatial scale $L = (C/\beta)^{1/2}$, the basic equations without external forcing and dissipation terms will be

$$u_t - yv + p_x = 0, \tag{5-19a}$$

$$v_t + yu + p_y = 0, \tag{5-19b}$$

$$p_t + \left(u_x + v_y\right) = 0. \tag{5-19c}$$

Now consider waves propagating in the east-west direction $A(y)\exp[i(kx - \omega t)]$. The above equations will reduce to

$$v_{yy} + \left(\omega^2 - k^2 + \frac{k}{\omega} - y^2\right)v = 0 \tag{5-20}$$

which is an Hermite differential equation. Since we are interested in the equatorial region, we use the boundary condition $v \to 0$ as $y \to \pm\infty$. Then the eigenfunctions should satisfy the following relation

$$\omega^2 - k^2 - \frac{k}{\omega} = 2n+1 \qquad n = 0,1,2,\ldots. \qquad (5\text{-}21)$$

The solutions for each n will be

$$v_n = H_n(y)\exp\left(-\frac{1}{2}y^2\right)\exp\left[i(kx - \omega t)\right]. \qquad (5\text{-}22)$$

H_n is the n-th Hermite polynomial. u and p can be obtained by substituting Eq. (5-22) into Eq. (5-19):

$$u_n = \frac{1}{i(\omega^2 - k^2)}\left(-\omega y v + k\frac{dv}{dy}\right) \qquad (5\text{-}23a)$$

$$p_n = \frac{1}{-i(\omega^2 - k^2)}\left(kyv - \omega\frac{dv}{dy}\right). \qquad (5\text{-}23b)$$

In addition to the above solutions, there is a solution with $v = 0$ in Eq. (5-19). This particular solution satisfying the boundary condition at infinity is: $\omega = k$ and

$$u = p = Ce^{-(1/2)y^2}\ \exp\left[i(kx - \omega t)\right]. \qquad (5\text{-}24)$$

Since substituting $n = -1$ into Eq. (5-21) will lead to $\omega = k$, we have now the complete set of solutions if we add this solution to Eq. (5-23) (Matsuno, 1966).

The dispersion diagram Fig. 5-4 shows the relation between the wave number and the phase velocity. $n = -1$ represents the Kelvin wave propagating eastward without dispersion and $n = 0$ corresponds to the mixed gravity–Rossby wave (Yanai–Maruyama wave). The mode n equal to or greater than 1 indicates inertial waves and equatorial Rossby waves. The former can propagate both eastward and westward while the latter propagate only westward very slowly. It is worth noting that the Rossby waves in the long wave region are almost non-dispersive and their group velocity is westward, whereas in the short wave region they are quite dispersive and their group velocity is eastward. The structures of the Kelvin wave and the first Rossby wave, which play a key role in the El Niño phenomenon, are shown in Fig. 5-5. The positive and negative anomalies in the figure correspond to the deep and shallow thermoclines, respectively. The amplitude of the Kelvin wave is the largest at the equator and the current velocity is eastward at the deep thermocline. The equatorial Rossby waves have their maximum pressure anomalies a little distance away from the equator, and the farther from the equator, the more parallel to the isobars their

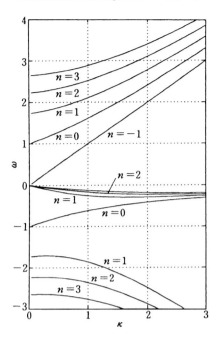

Fig. 5-4. Dispersion relation of equatorial waves. $n = -1$ indicates eastward propagating Kelvin wave, $n = 0$ is the mixed Rossby gravity wave, $n = 1, 2, 3, \ldots$ with small negative ω are Rossby waves, and $n = 1, 2, 3, \ldots$ with large positive (negative) ω are eastward (westward) propagating inertial gravity waves.

currents become, balancing in geostrophy. These waves are trapped near the equator. For instance, if C_m is 2.4 (m/sec), which is a typical lowest baroclinic mode in the Pacific, the spatial scale will be 325 km. The higher the modes are, the more trapped the waves are, as was indicated in Table 5-1.

5.2.3 Generation of equatorial waves by wind

In the previous section it was shown that the equator acts as if it were a wave guide, making waves trapped near the equator. These waves are very important for considering large-scale circulations and temporal variations in tropical oceans. In this section we will discuss how they are generated in the ocean.

a) Equation set with long wave approximation

For a phenomenon like El Niño, which evolves very slowly and is elongated in the east-west direction, inertia-gravity waves or short Rossby waves are not important. To eliminate these modes, we ignore the temporal derivative of the north-south velocity in Eq. (5-19), adopting the "long wave approximation". Then the equations will be

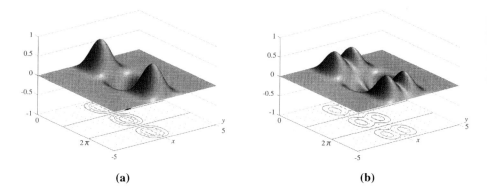

Fig. 5-5. Pressure pattern of (a) Kelvin wave ($n = -1$) and (b) the first symmetric Rossby wave ($n = 1$) for wave number $k = 1$. Positive (negative) values indicate deepening (rising) of the thermocline. The horizontal scale is normalized by $(\beta C)^{-1/2}$ (about 325 km for $C = 2.4$ m/s). The area between ± 5 is depicted in north-south and 3π in east-west.

$$u_t - yv + p_x = -\varepsilon u + F \,, \qquad (5\text{-}25\text{a})$$

$$yu + p_y = 0 \,, \qquad (5\text{-}25\text{b})$$

$$p_t + \left(u_x + v_y\right) = -\gamma p \,, \qquad (5\text{-}25\text{c})$$

where $\varepsilon = A\beta^{-1/2}C^{-5/2}$ and $\gamma = B\beta^{-1/2}C^{-5/2}$ are the dimensionless Rayleigh friction and the Newtonian cooling coefficient, respectively. $F = F^x(C\beta)^{-1/2}$ is a dimensionless external force due to wind stress.

To solve the above equations, it is rather convenient to introduce new variables, $s = p + u$ and $r = p - u$, following Gill (1980). For simplicity of calculation, we assume $\varepsilon = \gamma = a$, and expand each variable by means of the Hermite function Ψ_n (a complete orthogonal system)[4]:

---

[4]Ψ_n is defined using the Hermite polynomials as follows:

$$\Psi_n(y) = \pi^{-1/4}(2^n n!)H_n(y)\exp(-y^2), \qquad n = 0, 1, 2, \dots.$$

Ψ_n has the following orthogonal relation

$$\int_{-\infty}^{+\infty}\Psi_m \Psi_n dy = \delta_{mn}$$

where δ_{mn} is the Kronecker's delta (see Chapter 1).

$$(s, r, v, F) = \sum_{n=0}^{\infty} \left(s_n(t, x), r_n(t, x), v_n(t, x), F_n(t, x) \right) \Psi_n(y). \qquad (5\text{-}26)$$

Ψ_n has the recurrence formulae

$$\Psi_{ny} - y\Psi_n = -(2n+2)^{1/2} \Psi_{n+1}, \qquad (5\text{-}27a)$$

$$\Psi_{ny} + y\Psi_n = -(2n)^{1/2} \Psi_{n-1}. \qquad (5\text{-}27b)$$

Calculating the sum and difference of Eqs. (5-25a) and (5-25c), with the orthogonal formula of Ψ_n, we obtain

$$\frac{\partial s_n}{\partial t} + \frac{\partial s_n}{\partial x} + as_n - (2n)^{1/2} v_{n-1} = F_n, \qquad (5\text{-}28a)$$

$$\frac{\partial r_n}{\partial t} + \frac{\partial r_n}{\partial x} + ar_n + (2n+2)^{1/2} v_{n+1} = -F_n. \qquad (5\text{-}28b)$$

Eq. (5-25b) will provide the relation between s and r:

$$r_n = \left(\frac{n+2}{n+1} \right)^{1/2} s_{n+2}. \qquad (5\text{-}29)$$

Eliminating v_n leads to, for n equal to or greater than 1,

$$\frac{\partial s_n}{\partial t} + \frac{1}{1-2n} \frac{\partial s_n}{\partial x} + as_n = \frac{n-1}{2n-1} F_n - \frac{(n^2-n)^{1/2}}{2n-1} F_{n-2} \qquad (5\text{-}30)$$

and for $n = 0$,

$$\frac{\partial s_0}{\partial t} + \frac{\partial s_0}{\partial x} + as_0 = F_0. \qquad (5\text{-}31)$$

The above equations are wave equations with forcing terms. The case $n = 0$ represents the Kelvin wave and others the Rossby waves. The odd modes ($n = 1$, 3, 5, …) are antisymmetric about the equator while the even modes ($n = 0, 2, 4,$ …) are symmetric. The value of s_n determines the amplitude and the phase of each mode. Under the long wave approximation the phase speed of the equatorial Kelvin wave is 1 (C_n in dimensional value) and that of the Rossby wave is $1/(1 - 2n)$ ($C_n/(1 - 2n)$ in dimensional value).

b) *Eastern and western boundaries*
    Appropriate boundary conditions are necessary to solve Eqs. (5-30) and (5-31). In the case of the Pacific, setting walls at the eastern end ($x = x_E$) and western end ($x = 0$) will be sufficient for the first approximation. At the eastern end no flow across the wall can be easily justified. However, to satisfy the boundary condition $u = 0$ at the western end, we should take into account the short Rossby waves, which were omitted under the long wave approximation. Therefore, we need to find the right boundary condition for this approximate system. The short Rossby waves form the western boundary layer, without penetrating into the wall. Outside the boundary layer, the net east-west velocity can be shown to be zero when integrated in the north-south direction (Cane and Sarachik, 1977). This is a very natural condition to make the total east-west mass transport be zero if integrated meridionally. Then, we can impose

$$\int_{-\infty}^{\infty} u \, dy = 0 \qquad x = 0, \qquad (5\text{-}32a)$$

$$u = 0 \qquad x = x_E. \qquad (5\text{-}32b)$$

The eastern boundary condition can be rewritten using the orthogonal relation of Ψ_n:

$$s_n(x_E) = \left[\left(\frac{n-1}{n}\right) \cdot \left(\frac{n-3}{n-2}\right) \cdots \frac{1}{2}\right]^{1/2} s_0(x_E) \qquad n = 2, 4, 6, \ldots. \qquad (5\text{-}33)$$

This means that an equatorial Kelvin wave that has reached the eastern boundary is reflected as Rossby waves whose ratios are given by s_n in Eq. (5-33). Since the Kelvin wave is symmetric about the equator, no antisymmetric Rossby waves will be reflected from a straight rigid wall along a meridian.
    At the western boundary, the recurrence relationship of Ψ_n, that is,

$$\int_{-\infty}^{\infty} \Psi_n \, dy = \left(\frac{n-1}{n}\right)^{1/2} \int_{-\infty}^{\infty} \Psi_{n-2} \, dy \qquad (5\text{-}34)$$

can be used to obtain

$$s_0(0) = \sum_{n=2,4,6,\ldots}^{\infty} \left[\left(\frac{n-1}{n}\right) \cdot \left(\frac{n-3}{n-2}\right) \cdots \frac{1}{2}\right]^{1/2} \cdot (n-1)^{-1} \cdot s_n(0). \qquad (5\text{-}35)$$

This indicates that Rossby waves that have reached the western boundary are summed up in the ratios as shown on the right-hand side of Eq. (5-35) and reflected as an equatorial Kelvin wave. If the amplitude of each Rossby wave is 1, the ratios of its contribution to the Kelvin wave are, 0.71 (for $n = 2$), 0.20 ($n = 4$), and 0.11 ($n = 6$). When Rossby waves of lower modes enter into the western boundary in the equatorial region, the Kelvin wave is more effectively generated as a result of their reflection.

c) Response to uniform easterlies (trade winds)

The tropical Pacific is unique in that the thermocline is as deep as 150 m in the western region, but shallower than 50 m in the eastern region (Fig. 5-6). Since warm and light seawater is more abundant in the western Pacific the sea level is higher there than in the eastern Pacific by several tens of centimeters. Along the equator, a current with a speed of about 1 (m/s) and a width of a few hundreds of kilometers flows westward along the equator. It is called the South Equatorial Current (Fig. 5-7). A similar current exists in the Atlantic, but in the Indian Ocean a similar but eastward flow occurs twice a year in the monsoon transition seasons, namely, in May and October. This flow is called the Wyrtki jet (Yamagata and Matsuura, 1993). In theory such flows are generically called the Yoshida jet (Yoshida, 1959).

How is the Yoshida jet generated? How is the east-west slope of the thermocline or the pressure gradient maintained? A key to solve these problems lies in the trade winds blowing westward in the tropics. Here we will examine what kind of oceanic circulation is formed in response to uniform trade winds. If we assume the external force $F = 1$, then

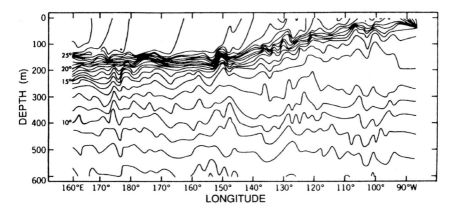

Fig. 5-6. Vertical cross section of temperature along the equator.

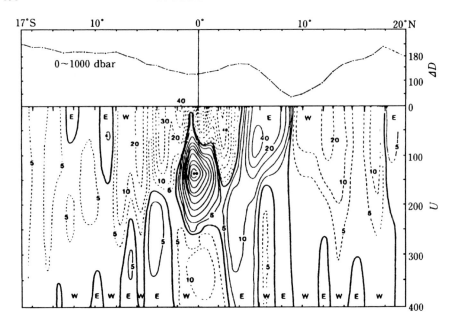

Fig. 5-7. Meridional vertical cross section of east-west velocity and dynamic height (Wyrtki and Kilonsky, 1984). The dynamic height (ΔD) is calculated from the reference level of 1000 dbar. The data is based on the ship observations between Hawaii and Tahiti in 1979–1980.

$$F = 1 = \sum_{n=0,2,4,\ldots} F_n \Psi_n . \qquad (5\text{-}36)$$

Using Eq. (5-27) and the orthogonal relation of Ψ_n, we obtain

$$F_n = \int_{-\infty}^{\infty} \Psi_n dy = \left(\frac{n-1}{n}\right)^{1/2} F_{n-2} = \frac{2^{1/2}\pi^{1/4}}{2^{n/2}(n/2)!}\sqrt{n!} . \qquad (5\text{-}37)$$

After substituting this external forcing term, Eqs. (5-30) and (5-31) will be solved. First we assume that the solutions are stationary and uniform in the east-west direction without any boundaries. In this case derivatives with respect to x and t can be replaced by zero, so that s_n can be easily obtained. The solutions are

$$u(y) = \frac{-2}{a} \sum_{n=0,2,4,\ldots}^{\infty} \frac{F_n}{(2n-1)(2n+3)} \Psi_n(y) , \qquad (5\text{-}38a)$$

$$v(y) = \sqrt{2} \sum_{n=0,2,4,\ldots}^{\infty} \frac{(n+1)F_n}{(2n+3)} \Psi_n(y),$$ (5-38b)

$$p(y) = \frac{-2}{a^2} \sum_{n=0,2,4,\ldots}^{\infty} \frac{(2n+1)F_n}{(2n-1)(2n+3)} \Psi_n(y).$$ (5-38c)

As is seen from Fig. 5-8, the current is concentrated in a narrow band near the equator, the width is of an order of the equatorial radius of deformation.

If the boundary conditions Eqs. (5-33) and (5-35) are included, the solutions bounded by two walls are found after complicated calculations. They are shown in Fig. 5-9 (for analytical solutions, see Yamagata and Philander, 1985). The characteristic of the equatorial ocean is well described in the figure: the eastward jet is localized in the equator region, and the thermocline is deep in the west and shallow in the east. Thus the oceanic circulation in the equatorial region is accounted for mainly as a response to the trade winds.

So far, the vertical eddy viscosity of momentum and the vertical diffusion of heat have been assumed to be equal. When this assumption is not adopted, the following scale transformations can be made for all the terms in Eq. (5-25) except the temporal derivative terms.

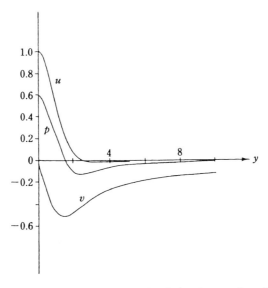

Fig. 5-8. Meridional structure of the ocean circulation (u, v, p) forced by a spatially homogeneous easterly (Moore and Philander, 1977).

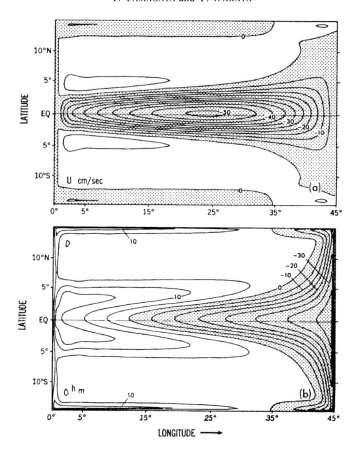

Fig. 5-9. Circulation pattern of the ocean, bounded by eastern and western rigid walls, forced by a spatially homogeneous easterly. (a) East-west velocity. The contour interval is 5 m, and the westward flow is shaded. (b) Thermocline depth. The contour interval is 5 cm/s, and the shallower thermocline is shaded.

$$
\begin{bmatrix} x \\ y \\ u \\ v \\ p \\ F \end{bmatrix} = \begin{bmatrix} (\varepsilon/\gamma)^{1/2}\,X \\ (\varepsilon/\gamma)^{1/4}\,Y \\ (\gamma/\varepsilon)^{1/2}\,U \\ (\gamma/\varepsilon)^{3/4}\,V \\ P \\ (\gamma/\varepsilon)^{1/2}\,F \end{bmatrix}. \tag{5-39}
$$

Then the system of equations will have the same form as when $\varepsilon = \gamma$. This formula also indicates that the width of the Yoshida jet is proportional to $(\varepsilon/\gamma)^{1/4}$.

When we consider a stationary state, the effect of dissipation is quite important. Let us see how important the vertical thermal diffusion is for accomplishing a stationary state. As $\gamma \to 0$, we can obtain from Eq. (5-25):

$$u = v = 0, \qquad p_x = -1. \tag{5-40}$$

That is, a state of no motion appears. In other words, the external force is exactly in balance with the pressure gradient. On the other hand when $\varepsilon \to 0$, the intensity of the Yoshida jet will be infinitely large, with an infinitesimally narrow width.

d) Response to westerlies in a limited region

It is known from the previous discussions that the trade winds generate the surface circulation in the equatorial ocean. At the onset of El Niño, an anomaly of westerlies takes place west of the International Date Line in the western equatorial Pacific, which in the development stage may reverse the direction of the trade winds in the basic field. Let us suppose westerlies are blowing in a limited area near the equator and examine the response with special attention to the Kelvin wave and the first symmetric Rossby wave.

Assume that the wind is blowing eastward at 2 m/sec over an area extending 5000 km in the east-west direction within 10 degrees of latitudes:

$$U = 2 \ (\text{m s}^{-1}) \exp\left[-\frac{y^\circ}{10} \right]^2. \tag{5-41}$$

For nondimensionalization, we use the phase speed 2.4 m/sec of a gravity wave in the ocean, the representative temporal scale 1.35×10^5 seconds, and the spatial scale 325 kilometers. The external force is expanded in Hermite functions as

$$F = F_0 \Psi_0 + F_2 \Psi_2 + \cdots. \tag{5-42}$$

If we choose for the wind stress parameter $(F = K_s U)$, $K_s = 0.02$ (its corresponding dimensional value will be 1.5×10^{-7}/sec), then $F_0 = 0.024$, $F_2 = 0.0113$, etc. are obtained.

When the dissipation is neglected, the solution for the Kelvin wave can be derived from

$$\frac{\partial s_0}{\partial t} + \frac{\partial s_0}{\partial x} = F_0 \tag{5-43}$$

and its equilibrium solution is

$$s_0 = \begin{cases} 0 & x < 0 \\ F_0 x & 0 \le x < L \\ F_0 L & L \le x. \end{cases} \qquad (5\text{-}44)$$

The Rossby wave solution can be obtained from

$$\frac{\partial s_2}{\partial t} - \frac{1}{3}\frac{\partial s_2}{\partial x} = \frac{1}{3}F_2 - \frac{\sqrt{2}}{3}F_0 \equiv F_R \qquad (5\text{-}45)$$

with its equilibrium solution

$$s_2 = \begin{cases} -3F_R L & x < 0 \\ 3F_R(x - L) & 0 \le x < L \\ 0 & L \le x. \end{cases} \qquad (5\text{-}46)$$

The depth anomaly $h = H[s_0 + \sqrt{2}\, s_2)\Psi_0 + s_2\Psi_n]/2 \cdot H$ (H is the representative vertical scale = 150 m) of the thermocline is estimated to be 19 m high at the eastern end and 13 m low at the western end of the wind blowing region. When the thermocline deepens in the east, the information of the descending area is transmitted to the east as a Kelvin wave. Meanwhile the thermocline rises in the west and its information propagates to the west as Rossby waves.

e) *Effects of winds oscillating with time*
       Corresponding to El Niño and La Niña (the opposite phase of El Niño), we examine the response of the closed ocean with a width of 15,000 km to the east-west wind field oscillating with a period of three years. The viscous effect is assumed to be $a = (2.5 \text{ years})^{-1}$.

$$U = 2\,(\text{m s}^{-1})\exp\left[-\left(\frac{y^\circ}{10}\right)^2 - \left(\frac{x - 21}{9}\right)^2\right]\sin\left[\frac{2\pi}{(3 \text{ years})}t\right]. \qquad (5\text{-}47)$$

Figure 5-10 shows the variation of the thermocline depth at the equator for 2000 days. The Kelvin and Rossby waves are emitted to the east and the west, respectively, and they reflect at the eastern and western boundaries. It is seen that the thermocline vacillates in the east-west direction as if it were a seesaw.

5.2.4 *Thermodynamics of the oceanic mixed layer*
       The equation that determines the temperature anomaly can be expressed, after the thermodynamic equation is linearized, by

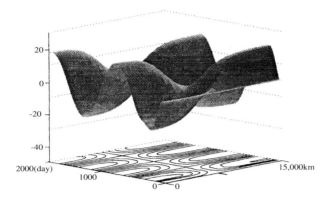

Fig. 5-10. Thermocline depth anomaly along the equator, in the case that the east-west winds are oscillated with a period of three years. This is a dimensional value (m), and the temporal variabilities are depicted for 200 days.

$$\frac{\partial T}{\partial t} = -u\overline{T}_x - \left(\text{cold water entrainment by upwelling}\right) - n_d T . \qquad (5\text{-}48)$$

The first term on the right-hand side represents anomalous advection on the mean zonal sea surface temperature gradient, the last term Newtonian cooling due to thermal radiation. Zebiak and Cane (1987) introduced a boundary layer of depth $H_1 = 50$ m into the surface of the mixed layer and derived the equation to forecast its temperature. Following their method, we are going to estimate the vertical mixing term. After the second term of upwelling entrainment is rewritten as a difference between the temperatures of the sea surface and that at 50 m deep, divided by the depth H_1, the above equation will be

$$\frac{\partial T}{\partial t} = -u\overline{T}_x - \gamma\overline{W}_s \frac{\left(T - T_s\right)}{H_1} - n_d T . \qquad (5\text{-}49)$$

Here $\overline{W}_s$ is the averaged upwelling, and γ (~0.75) the mixing ratio of water entrained from the lower layer. In other words, the second term represents the effect of cooling by the lower layer water with the temperature T_s entering into the surface layer. The temperature anomaly in the lower part of the mixed layer is high when the thermocline is deep; it is low when the thermocline is shallow. Hence, a typical form of T_s can be represented by

$$T_s = A\left\{\tanh\left[B\left(\overline{h} + |h|\right)\right] - \tanh\left(B\overline{h}\right)\right\} \qquad (5\text{-}50)$$

where h is the fluctuation (anomaly) to the averaged depth of the mixed layer.

A and B are constants: for $h \geq 0$, $A = 28$ K, $B = 1/(80$ m$)$; for $h < 0$, $A = -40$ K, $B = 1/(33$ m$)$. Expanding T_s into Taylor series and neglecting the second or higher order terms, we obtain

$$\frac{\partial T}{\partial t} = -u\overline{T}_x + \frac{\gamma\overline{W}_s}{H_1}\left(\frac{AB}{\cosh^2(b\overline{h})}\right)h - \left(\frac{\gamma\overline{W}_s}{H_1} + n_d\right)T \qquad (5\text{-}51)$$

which can be simplified to

$$\frac{\partial T}{\partial t} = -u\overline{T}_x + K_T h - dT \qquad (5\text{-}52)$$

where $d \equiv (\overline{W}_s/H_1 + n_d)$, $K_T \equiv (\overline{W}_s \gamma/H_1) < AB\cos h^{-2}B\,\overline{h} >$ ($<AB\cos h^{-2}B\,\overline{h}>$ is the mean value for the cases of positive and negative h).

   If the typical east-west temperature gradient is $\overline{T}_x = -5 \times 10^{-7}$/km, then the temperature difference across the Pacific will be about 8°C. For the mean depth $\overline{h} = 50$ m and $\overline{W} = 2$ m/day, K_T will be 8×10^{-8}/km/s and d will be 1/26 days.

5.3 ATMOSPHERIC MODELS IN THE TROPICS

   Evaporation is quite large in the tropics, which results in high cumuli. Latent heat of evaporation is emitted to the atmosphere through rainfall. Consequently, the atmosphere is heated. Here we will discuss the response of the heated atmosphere using a simple model. Since sea surface temperatures vary very slowly compared with atmospheric phenomena, the atmosphere can be considered in a quasi-equilibrium state. Hence we will not include the direct temporal variation in the atmospheric model. We are interested in the interaction with the ocean, so that we pay attention to the winds in the lower boundary layer of the atmosphere, and make a simple model taking transports in the boundary layer as dependent variables:

$$U = \int_{p_S}^{p_T} \frac{u\,dp}{g} \qquad (5\text{-}53)$$

$$V = \int_{p_S}^{p_T} \frac{v\,dp}{g} \qquad (5\text{-}54)$$

where p_s is the pressure at the sea surface, p_T the pressure at the upper end of the boundary layer. Pressure is used as the vertical coordinate. Linearized equations of motion and continuity can be written as (Holton, 1975; Philander, 1990)

$$A_* U - fV + \varphi_x = 0, \tag{5-55a}$$

$$A_* V + fU + \varphi_y = 0, \tag{5-55b}$$

$$\omega_* + U_x + V_y = 0, \tag{5-55c}$$

where φ is the geopotential, A_* the Rayleigh friction, ω_* the upward velocity dp/dt defined in the pressure coordinate. In the tropics where the Coriolis force is small, heating directly causes upward motion in the atmosphere. If we assume that the Newtonian cooling is proportional to φ, Eq. (5-55c) can be rewritten as

$$B_* \varphi + C_A \left(U_x + V_y \right) = -Q \tag{5-56}$$

where Q is the heating function due to precipitation. C_A indicates how much the atmosphere is stable and it does not represent the phase speed of the gravity wave in the atmosphere. The typical values are: $C_A = 60$ m/sec, $B_* = 1/(2 \text{ days})$. We now have exactly the same type of equation as for the ocean. However, it should be noted that the parametric values for the atmosphere are very different from those for the oceanic model ($C = 2.4$ m/sec, the dissipation rate is about several years). This set of equations governing the forced motion of the atmosphere was first derived by Matsuno (1966), who obtained the response to periodic forcing in the east-west direction.

Gill (1980) obtained an analytical solution for the response to a locally confined forcing that is symmetric about the equator (Fig. 5-11). In this case stationary decaying Rossby waves are generated on the western side of the heat source, and the stationary decaying Kelvin wave on the eastern side. Heating does not add net momentum to the system and consequently the westerlies associated with the Rossby wave are three times as strong as the easterlies associated with the Kelvin wave on the equator. This is because the group velocity of the Rossby wave is one third of that of the Kelvin wave, which limits the extent of the westerlies within one third of that of the Kelvin wave.

5.4 OCEAN–ATMOSPHERE COUPLING PROCESSES

So far, we discussed the ocean and the atmosphere separately: the oceanic circulation in the tropics was considered as a response to winds, and the atmospheric circulation as a response to heating. In this section we will discuss the coupling processes of the ocean and the atmosphere.

5.4.1 Ocean–atmosphere coupling

The wind blowing over the sea surface transfers momentum to the ocean through friction to generate motion in the ocean. The wind stress is given by

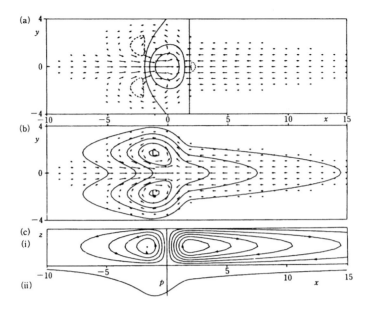

Fig. 5-11. Atmospheric response to the heating north-south symmetrical against the equator. (a) Vertical velocity, solid lines indicate upward velocities and broken lines downward velocities. Strong upward velocity area corresponds to the location of the heating area. Vectors are the low level velocities. (b) Pressure pattern. (c) Vertical east-west cross section along the equator of (i) stream function and (ii) pressure anomaly integrated meridionally.

$$\left(F^x, F^y\right) = K_S(U, V) \tag{5-57}$$

where K_S is the coefficient of proportionality. If we assume that much evaporation is produced over warm seawater and cumuli are generated, emitting latent heat through rainfall, then the atmospheric heating function Q can be roughly approximated by

$$Q = K_Q T \tag{5-58}$$

where T is the sea surface temperature, K_Q the coefficient of proportionality. This will allow us to construct a coupled model as follows:

For the ocean,

$$u_t - fv = -g^* h_x + K_S U - au, \tag{5-59a}$$

$$v_t + fu = -g^*h_y + K_S V - av \,, \tag{5-59b}$$

$$h_t + H\left(u_x + V_y\right) = -bh \,, \tag{5-59c}$$

$$T_t = -uT_x + K_T h - dT \,. \tag{5-59d}$$

For the atmosphere,

$$A_* U - fV + \varphi_x = 0 \,, \tag{5-59e}$$

$$A_* V + fU + \varphi_y = 0 \,, \tag{5-59f}$$

$$B_* \varphi + C_A\left(U_x + V_y\right) = -K_Q T \,. \tag{5-59g}$$

In the above oceanic part of the mode, the anomaly h of the mixed layer is used in place of p. As described for Eq. (5-10), the conversions $p \to gh$ and $C_m^2 \to gH$ will result in the above equations. Since H is determined from the eigenvalue C_m^2, its value (see Table 5-1) would be very different from the real thickness of the mixed layer. Thus, sometimes H is assumed to have a fixed value and g is taken as a variable. For instance, if we assume H to be 150 m, the representative thickness of the mixed layer, g ($= C_m^2/H$) would be much smaller than the gravitational acceleration ($= 9.8$ m/sec^2). Then, it will be expressed as g^* (reduced gravity) to be distinguished from g. In Eq. (5-59) where g is replaced by g^*, H will be the thickness of the mixing layer in the mean field, and h the deviation from H[5].

5.4.2 Coupled instability of ocean and atmosphere

As people often say "which is first, egg or chicken?"; some argue whether the atmosphere causes oceanic motion or the ocean causes the atmospheric motion. In fact the ocean and atmosphere interact with each other and a united phenomenon evolves in time like a self-excited oscillation. Here let us make a thought experiment on the ocean–atmosphere-coupled instability. Firstly, as

---

[5] Concerning the thickness of the mixing layer, a misinterpretation is sometimes found in literature. It should be noted that H is determined by an eigenvalue that corresponds to a particular vertical mode and which vertical mode is predominant is associated with the spatial and temporal scale of the external force. For example, for a slowly varying external force as in the case of El Niño, the second vertical mode ($C_2 \sim 1.4$ m/sec) will dominate, while for the scale of the westerly bursts that accompany the seasonal disturbances in the atmosphere, the first vertical mode ($C_1 \sim 2.4$ m/sec) will be resonantly excited. Even if we use the above mentioned concept of reduced gravity, no mixing layer of such thickness exists in reality.

shown in Fig. 5-12, over a high temperature area where the mixed layer is thicker with the depressed thermocline, seawater tends to evaporate more and cumulus clouds are generated. Condensation in the clouds causes a heating of the atmosphere, resulting in the upward motion accompanied by the wind blowing into the area. This convergent wind transfers its momentum to the ocean accelerating the converging current. Thus the initial depression of the thermocline grows, suppressing the upwelling of cold water from below. As a result the sea surface temperature rises, which in turn causes convergent wind. This is nothing but a positive feedback mechanism, which is built into the ocean–atmosphere system, indicating the existence of an unstable solution in Eq. (5-59). From this thought experiment, it is anticipated that a necessary condition for instability to occur is a positive correlation between the wind and the current on the average. In fact, the model energy equation derived from Eqs. (5-59a), (5-59b) and (5-59c) will be

$$\frac{1}{2}\left\langle H\left(u^2+v^2\right)+g^*h^2\right\rangle_t = -a\left\langle H\left(u^2+v^2\right)\right\rangle - b\left\langle g^*h^2\right\rangle + K_S\left\langle H(uU+vV)\right\rangle$$

$$(5\text{-}60)$$

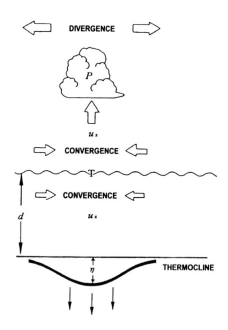

Fig. 5-12. Schematic diagram of an instability in the tropics, which occurs by the coupling of atmosphere and ocean. Deepening of the thermocline (η) induces the warming of sea surface temperature (T), and hence the atmospheric heating (P) is enhanced, and lower winds converge, which result in the convergence of ocean currents. Finally η increases furthermore.

where < > denotes integration over the whole domain. The first and second terms on the right-hand side are negative definite, so that for an instability to occur, the third term must be positive. Namely, the necessary condition for instability is that atmospheric winds and oceanic flows have a positive correlation in total (Yamagata, 1985). Adiabatic wave dynamics in the ocean plays a dominant role in the evolution equation for the El Niño, and the non-adiabatic process in the ocean and the atmosphere follows as a quasi-stationary state. This is the reason why the standard dynamic equation (5-60) can provide the necessary conditions for coupled instability. In the case of the ocean–atmosphere coupled system in mid- and high-latitudes, a more local and non-adiabatic process in the ocean will become rather important, adding complexities to the problem.

5.4.3 Eigenvalue problem for coupled instability

There are various types of coupled instability depending on different factors that determine sea surface temperature. Hirst (1986) solved an eigenvalue problem by assuming a solution to be of a periodic wave type in the east-west direction, $A(y)\exp[i(kx - \sigma t)]$. The imaginary part of σ represents the growth rate, the real part the frequency. A few examples will be given below.

a) Model I: Local thermal equilibrium limit

We will consider a case where an upwelling effect is strong and the third and fourth terms in the thermodynamic equation (5-59d) are significant. In this case, the thermodynamic equation will be simplified as follows (Philander et al., 1983; Yamagata, 1985).

$$T = \kappa h \qquad (5\text{-}61)$$

where $\kappa = K_T/d$. As can be understood from the definition of K_T this limit holds for the oceanic region for small $\bar{h}$, that is, where the mixed layer is thin[6]. The eigenvalue σ versus the wave number k is plotted in Fig. 5-13. It is seen that the Kelvin wave in the ocean is deformed because of the interaction with the atmosphere and becomes unstable in the long-wave region. It is stable in the short-wave region. This is because the buoyancy force due to stratification increases as the wavelength becomes shorter, suppressing the coupled instability. Its phase speed is slightly slower than that of the pure oceanic Kelvin wave. The structure of the unstable solutions is shown in Fig. 5-14(A). The oceanic Kelvin wave has large current velocity when the mixed layer is

---

[6]Nonetheless, this proportionality relationship empirically holds for the areas of thick thermoclines in the western and central Pacific. The proportionality constant κ is small but a slight rise in the seawater temperature will greatly affect the cumulus activities in the atmosphere in these areas of high sea surface temperatures. In this sense they are important areas for the ocean–atmosphere couples system.

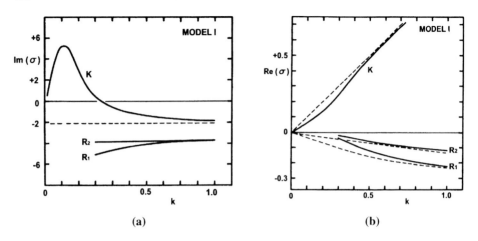

Fig. 5-13. Nondimensional growth rate (Im(σ)) and frequency (Re(σ)) vs wave number k for model I. Im(σ) is drawn as 10^{-2} is 1 unit, and the unit of Re(σ) is 1. $k = 0.1$ corresponds to 16,000 km. Broken lines indicate the case of non-coupling between atmosphere and ocean. K is the Kelvin wave, and R_1 and R_2 are Rossby waves.

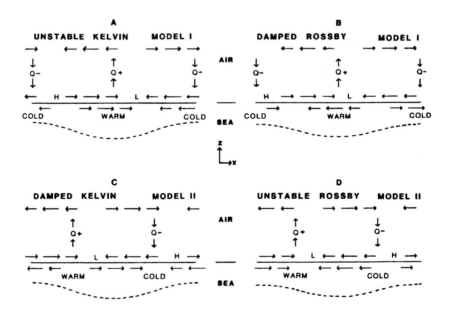

Fig. 5-14. Schematic diagrams of motions along the equator, which correspond to the Kelvin wave and Rossby wave for Model I ((A), (B)) and Model II ((C), (D)), respectively (Hirst, 1986). The solid line indicates the sea surface level, and the broken line is the thermocline depth. The arrows indicate oceanic currents and atmospheric motions. The maximum and minimum of sea surface temperature are expressed by warm and cold, respectively. The atmospheric heatings are Q^+ and Q^-, and the maximum and minimum pressure are H and L, respectively.

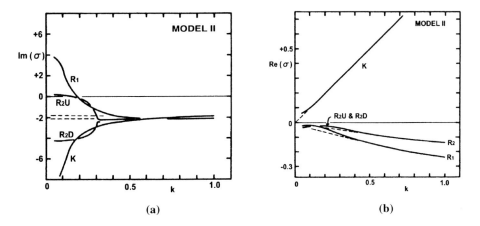

Fig. 5-15. Dispersion relation of Model II (Hirst, 1986). Inscription is the same as Fig. 5-13. Symbol U indicates unstable modes and D damped modes.

thick with a deep thermocline. At the locally thermal equilibrium limit the position of the deep thermocline coincides with the area of high sea surface temperature, and hence the area heating the atmosphere. Therefore, the strong westerlies accompanying the stationary Rossby waves in the atmosphere and the currents created by the oceanic Kelvin wave are both in the same eastward direction, so that the current is likely to be intensified further, that is, the necessary condition for instability is satisfied. On the other hand, oceanic Rossby waves attenuate even in the long wavelength range, and they will not become unstable. That is because at the high temperature region of the sea surface, the current flows westward, while the wind blows in the opposite direction, as is seen from Fig. 5-14(B).

b) Model II: Limit of horizontal advection

We consider the case when the east-west advection term is dominant in the thermodynamic equation that governs the sea surface temperature. Let us assume that there is no upwelling and $K_T \to 0$, then (Rennick, 1983; Gill, 1985)

$$T_t + \overline{T}_x u + dT = 0 . \qquad (5\text{-}62)$$

Figure 5-15 shows the growth rate and the frequency versus the wave number for each mode. The growth rate for the lowest symmetric mode Rossby wave (R_1) increases as the wavelength increases. The anti-symmetric Rossby wave (R_2) is divided into two modes, an unstable wave (R_2U) and a damping wave (R_2D). These unstable waves propagate westward more slowly than the free Rossby wave. As is seen from Fig. 5-14(D), the warm oceanic region in this case shifts to the west compared with Model I, and as a result the wind blows in the same direction as the current, satisfying the necessary instability condition. The Kelvin wave in this situation damps as shown in Fig. 5-14. This limit is applicable in the area with a large $\bar{h}$, i.e. in the western Pacific with deep thermocline, as found from the definition of K_T. The importance of the heat advection was first introduced to explain the evolution of ENSO by Gill (1985) and others. But today we know that the burst of the westerlies accompanying intra-seasonal variations generates an equatorial Kelvin wave and as the wave moves to the east "the abrupt response of the ocean" takes place.

c) Model III: General case

If all terms in the thermodynamic equation are included (Hirst, 1988), unstable stationary waves exist in addition to equatorial waves (Fig. 5-16). The eigenfunctions shown in Fig. 5-17 indicate that the mixed layer is the thickest in the equatorial area, similar to the Kelvin wave, but the maximum of the current is shifted westward in phase with the wind. The convergence zone of the current coincides with the deepest region, so that the wave becomes almost stationaly, since it is known from the continuity equation that the position of the deepest region should move towards the convergence zone of the current.

5.5 SIMPLIFIED OCEAN–ATMOSPHERE MODEL

So far we have discussed how El Niño can take place and develop from a viewpoint of the ocean–atmosphere coupled instability. To explain the cyclic nature of El Niño with intervals of 2–9 years, we need to understand the mechanism of the growth-development-decay process. This will lead to forecasting of El Niño and therefore an important theme from a societal point of view.

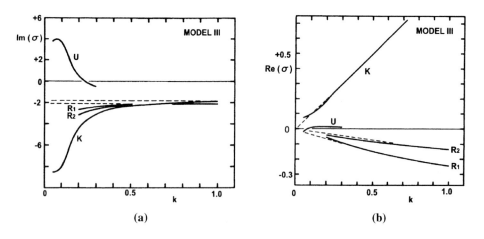

Fig. 5-16. Dispersion relation of Model III (Hirst, 1986). Inscription is the same as Fig. 5-13.

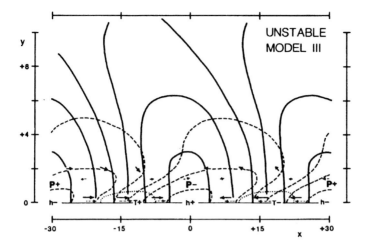

Fig. 5-17. Structure of the unstable eigenfunction in Model III (Hirst, 1986). Solid, broken, and dotted lines indicate lower atmospheric pressure (P), thermocline depth (h), and sea surface temperature (T), respectively. Solid and broken arrows indicate winds over the sea and ocean currents. The horizontal nondimensional 1 unit corresponds to 250 km.

We are going to introduce some time-evolution models which reproduce periodic oscillations and to clarify their mechanism. Most of the results of numerical calculations can be classified into two large groups. One is the model with eastward migration of the disturbance as shown in Model I in the previous section. The other is the model that exists like a stationary wave with no tendency of propagation. It might be thought that the difference between the two is only in appearance, but the fact is that the two models are essentially different in the oscillation mechanism.

5.5.1 Model of eastward propagating disturbances in the ocean and atmosphere

Anderson and McCreary (1985) first succeeded in reproducing the whole process of the growth, development and decaying of El Niño, the occurrence of La Niña, and the reoccurrence of the following El Niño. They set up an atmospheric model in the region extending in the east-west direction of $x = 0$ to 30,000 km, and an oceanic model within $15,000 < x < 30,000$ km region for solving the temporal evolution problem of the coupled model. Although their model is principally the same as the one discussed in the previous section, it is a nonlinear model, trying to obtain the solutions for the basic fields in addition to the fluctuations (anomalies).

Figure 5-18 shows time sections for 16 years of equatorial temperature, layer thickness and wind stress. After an initial period of adjustment the model settles down into an oscillatory mode with a period of about 3.5 years. Instabilities develop in the western ocean, propagate eastward slowly, and dissipate at the

(a)   **T**

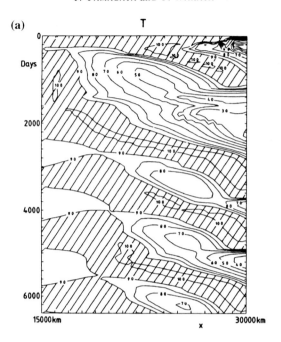

Fig. 5-18. Numerical simulation of Anderson and McCreary (1985). (a) Temporal variability of temperature along the equator (east-west size 15,000 km of the numerical model). (b) Thermocline depth; area deeper than 100 m is shaded. (c) Wind stress; positive area is shaded.

eastern boundary. The obtained pattern is similar to the result of the linear eigenvalue problem in Model I.

The mechanism of the oscillation mentioned above was further explored by Yamagata and Masumoto (1989; see also Masumoto and Yamagata, 1991). Let us consider the mechanism of oscillation. The anomaly propagates eastward and the ocean is bounded by east-west ends, hence some seed should be needed in the western Pacific to continue the oscillation. Suppose there is a warm region in the western Pacific. This anomaly propagates eastward as a coupled disturbance with its increasing amplitude. The western Pacific is eventually covered with westerlies blowing into this warm pool. The warm water in the mixed layer moves to the east due to the westerlies, and finally the western Pacific becomes cold, the eastern Pacific warm. This state itself will not create the subsequent El Niño. Heating over land plays an important role. Cooling in the western Pacific brings pressure difference in the atmosphere over land and sea: low over the continent and high over the ocean. This difference causes new easterly winds at the western end of the ocean. These winds will lead to a source of the next El Niño, making the western Pacific warm again. In fact, without the heat source over the continent, this model would not oscillate. If the heat source itself is made to

(b)   h

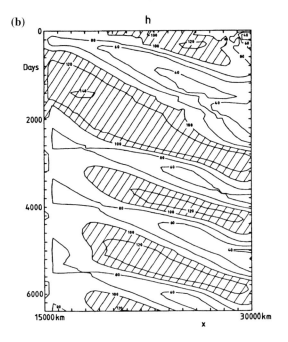

(c) τ^x

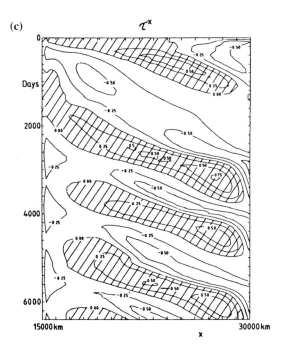

Fig. 5-18. (continued).

oscillate with a certain frequency, the coupled disturbance synchronizes to the frequency, which does not depend on the east-west scale of the Pacific (Masumoto and Yamagata, 1991). With such importance in the role of heat source over land, this model was the first to include "air–sea–land interaction" for the El Niño event.

It should be particularly noted that the solution of the above model is very similar to that observed in the biggest El Niño in 1982–1983. A good way of presenting the observational evidence on the relationship between SST and convection for the 1982–1983 event is by means of longitude–time plots of these quantities along the equator. The OLR (outgoing long wave radiation) anomaly is a good guide to the position of the anomalous region of convection, and Fig. 5-19(a) shows that this region, migrated slowly eastward, crossing the entire Pacific Ocean in the 16-month period depicted (Gill and Rasmusson, 1983). It is seen from Fig. 5-19(b) that the wind anomaly followed the OLR anomaly across the ocean. The SST pattern in Fig. 5-19(c) also showed a migration, as seen by following water over 29°C ("warm pool"). Comparing Fig. 5-19 with Fig. 5-18, we can tell that the model remarkably reproduces the characteristics of the real El Niño.

5.5.2 Model of standing oscillation

Zebiak and Cane (1987), by solving a simple ocean–atmosphere coupled model, obtained an ENSO-like oscillation. Their model, unlike the Anderson and McCreary model, forecasts fluctuations from the mean fields, or anomalies. In this sense, it can easily adjust the mean values to climatological data, which is a

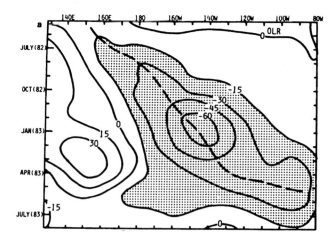

Fig. 5-19. Time–longitude section of the observation associated with the El Niño in 1982–1983 (Gill and Rasmusson, 1983). (a) Outgoing long-wave radiation (low value indicates active cloud). (b) Wind anomaly. (c) Sea surface temperature anomaly.

merit and demerit at the same time of this model. The atmospheric part of the model includes heating associated with convergence in the lower atmospheric layer in addition to heating proportional to sea surface temperatures. Figure 5-20(a) displays equatorial zonal wind stress anomalies between year 30 and year 45 of the 90-year simulation. (Positive (westerly) anomalies are indicated with solid lines, and negative (easterly) anomalies are indicated with dashed lines. Large westerly anomalies (>0.15 dyn/cm/cm) are stippled.) Figure 5-20(b) shows the model thermocline depth anomaly along the equator between year 30 and year 45 of the model simulation. (The contour interval is 10 m. Anomalies greater than 20 m are stippled; anomalies less than –20 m are hatched.) This variable may be

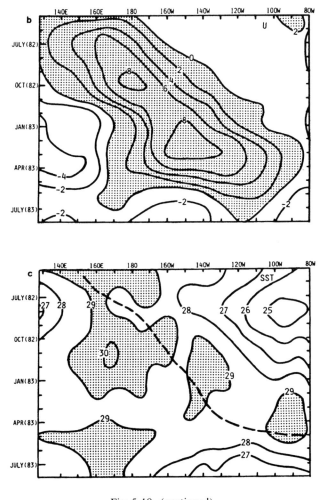

Fig. 5-19. (continued).

interpreted as a measure of the heat content of the upper ocean. The major warm episodes (beginning in years 31 and 41) are characterized by anomalously high heat content in the east and low heat content in the west for a period of nearly a year. This occurs approximately in phase with the strong and sustained westerly wind anomalies in the central Pacific (see Fig. 5-20(a)). Eastward propagation of the disturbances is not clearly seen compared with the result of Anderson and McCreary. The mixed layer depth oscillates in the east-west direction as if it were a seesaw. Such a standing oscillation is composed of a westward traveling and an eastward traveling waves. What are these waves? The mechanism can be

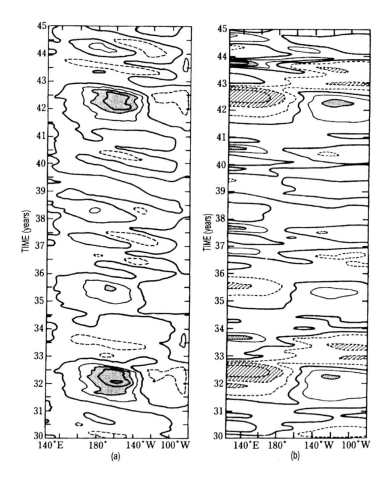

Fig. 5-20. Numerical simulation of Zebiak and Cane (1987). (a) Wind stress along the equator. Positive anomaly (westerly) is indicated by solid lines, and negative anomaly (easterly) is indicated by broken lines. Particularly, strong westerly anomaly (greater than 0.15 dyn cm^{-1}) is shaded. (b) Thermocline depth anomaly. The contour interval is 10 m. Larger than 20 m is shaded, and smaller than 20 m is shadowed.

explained by combining equatorial waves with the coupled ocean–atmosphere instability by a model proposed by Schopf and Suarez (1988; see also Suarez and Schopf, 1988) and Battisti (1988).

Their model is sketched in Fig. 5-21. It assumes the strongest coupling takes place in the central portion of the basin and that an important side effect of growing perturbations in this region is the emission of weakly coupled, westward propagating oceanic signals that, after reflecting from the western boundary, return and recouple to the atmosphere.

When a coupled instability occurs locally in the central region, wind anomalies appear there. If they are westerlies, a positive Kelvin wave will be emitted to the east, while negative Rossby waves will be emitted to the west as shown in Section 5.2.3 (positive and negative waves correspond to deep and shallow thermoclines, respectively). The positive Kelvin wave that has reached the eastern Pacific deepens the thermocline, preventing cold seawater from upwelling, which brings higher sea temperature. As a result convection and cumulus activities are enhanced over the warm sea and pressure is lowered due to the release of latent heat. Thus the wind blowing in the region becomes stronger and an El Niño develops. Meanwhile, the Rossby waves propagating westward will be reflected at the western wall and return as a negative Kelvin wave (see Fig. 5-22). It is noted that the reflected Kelvin wave has the opposite sign to that of the Kelvin wave created directly by the wind. Eventually the reflected Kelvin wave will reach the eastern Pacific and will shallow the thermocline; the positive anomaly will be changed to the negative one—the end of El Niño. Then over this cold seawater area cumuli disappear and wind weakens. The seawater temperature further lowers; namely a negative ocean–

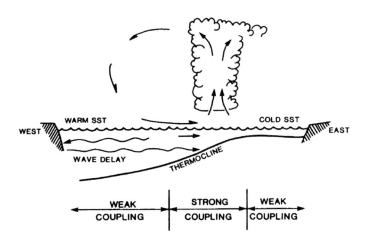

Fig. 5-21. Schematic diagram of the delayed oscillator mechanism (Suarez and Schopf, 1988).

El Niño

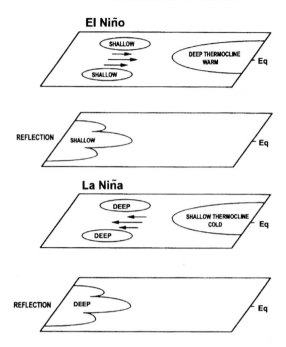

La Niña

Fig. 5-22. Schematic diagram viewing the role of equatorial waves on the delayed oscillator mechanism. Time progresses from top to bottom, and repeated.

atmosphere coupling occurs and negative temperature anomaly develops—the beginning of La Niña. This time, however, positive Rossby waves will be created and they will propagate to the west, and will be reflected at the western wall as a positive Kelvin wave.

This Kelvin wave will advance into the central portion of the Pacific and a series of processes will repeat. This is called the mechanism of "delayed oscillator" proposed by Suarez and Shopf (1988). This mechanism can be seen as an extension of Bjerknes's (1969) concept by adding the equatorial waves and their reflection at the western boundary.

Suarez and Shopf (1988) produced a similar model to Zebiak and Cane's. Figure 5-23 gives a synthesis of the El Niño cycle in a time–longitude domain to show the propagation of equatorial waves. Figure 5-23 (panel (a)) shows the thermocline depth anomalies averaged from 5°N to 7°N. This region slightly away from the equator is better to see the propagation of the Rossby wave since the amplitude of this wave has a maximum peak around there as shown in Fig. 5-23 (panel (b)). Note that the west is right and the east is left to see easily the reflection properties at the western boundary. The wave has large variability in the central and western Pacific. The panel (b) shows the depth anomalies but along the equator, and the west is left and the east is right. The propagation of the Kelvin wave can be seen in this figure, since this wave has a peak amplitude on

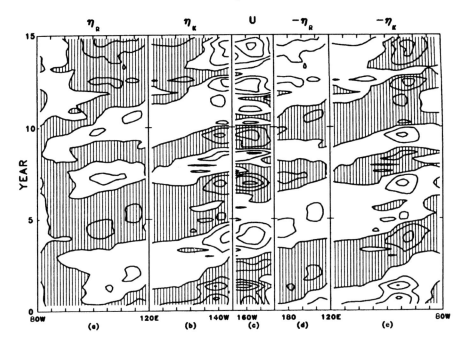

Fig. 5-23. Time–latitude cross section of the thermocline depth anomaly in the model of Schopf and Suarez (1988).

the equator. A tilting of phase toward the upper right in the combined two panels shows that the reflection of the Rossby wave at the western boundary produces the Kelvin wave. This wave is further amplified in the central Pacific. Since this growth has a good correlation with the east-west winds (panel (c)), it is speculated that the growth results from the atmosphere–ocean coupling instability. The panel (d) is the same as the panel (a), but only from the center of the wind anomaly out to the western boundary and the stippling is reversed to make easy to see the positive (westerly) wind (no stippling) anomaly producing the negative Rossby wave (no stippling in the panel (d)). The final panel (e) is a repetition of the panel (b) with reversed stripping, which completes the cycle.

The above results show that equatorial waves play the role of turnabout to the local coupling instability. The period of the cycle is far longer than the sum of the time of the Kelvin wave crossing the Pacific plus that of the Rossby wave crossing it. This is because we must take into account the growth time of the coupled instability to the time scale of the waves.

5.5.3 Interpretation using a delayed oscillator

The mechanism of the cyclic oscillation discussed in the previous section can be interpreted by a simplified model of "delayed oscillator equation"

(Suarez and Schopf, 1988; Battisti and Hirst, 1989). Let T be the seawater temperature anomaly in the eastern equatorial Pacific. Then the phenomenon is presumably governed by the equation

$$\frac{dT}{dt} = aT - bT(t - \tau) - cT^3 .$$

(5-63)

The first term on the right-hand side indicates linear instability due to ocean–atmosphere interactions. If we neglect the second and third terms, the solution will be $T \sim \hat{T}_0 \exp(at)$, that is, it will grow to e times as much as the initial value within $(1/a)$ time. If $\hat{T}_0$ is positive, this indicates that the high temperatures in the eastern Pacific create westerly wind components over the central Pacific, emitting Kelvin waves eastward, which further warm the eastern Pacific. If $\hat{T}_0$ is negative, on the other hand, this is a simplified representation of the process when lowering of seawater temperatures induce easterly wind components and further cool the eastern Pacific. In other words, the equation describes the coupled ocean–atmosphere instability in the eastern and central Pacific. The second term comes from the effect of the Kelvin wave generated by the reflection of Rossby waves. This suppresses the instability excited by the first term. The time τ is the "lag" time required for the Rossby wave to travel to the western boundary and come back to the central Pacific. The last cubic term is nonlinear and is a kind of the effect of autointoxication. That is, it reflects the effect that after the coupled instability grows to a certain level, the temperature of the mixed layer does not increase efficiently. Thus, the above simple equation contains all physical processes described earlier.

The characteristics of the equation can be known by the equilibrium solution (the solution of the equation without a time derivative) and its stability. For the case $b > a$ when the reflection parameter (b) is larger than the coupling parameter (a), there is one equilibrium solution:

$$T_0 = 0 .$$

(5-64)

If $a > b$, on the other hand, there are three equilibrium solutions:

$$T_0 = \begin{cases} 0 \\ \pm\sqrt{\dfrac{a-b}{c}} . \end{cases}$$

(5-65)

This means that there exist two possible cases for the solution in the eastern Pacific, warm and cold. We should examine the linear stability to tell whether or not such solutions are stable to be existent. This can be normally done by separating the solution into the equilibrium solution and a small perturbation of the order ε to linearize the equation and solve an eigenvalue problem.

$$T = T_0 + \varepsilon T' = T_0 + \varepsilon \tilde{T} e^{\sigma t} \qquad (5\text{-}66)$$

Substituting this into the delayed oscillator equation and taking the first order of ε, we obtain

$$i\sigma = a - be^{-\sigma t} - 3cT_0^2 \qquad (5\text{-}67)$$

Then we replace T_0 with the equilibrium solution Eq. (5-65) to examine its stability. The growth rate σ_r is given by the real part of σ. It is unstable when $\sigma_r > 0$, stable when $\sigma_r = 0$ or $\sigma_r < 0$. The imaginary part σ_i of σ represents the frequency. Figure 5-24(a) shows the instability of the solution $T_0 = 0$ for a reference value of $\tau = 180$ days, and $c = 0.07/K^2/year$, varying parameters (a, b). It is unstable for a broad region for the instability parameter a and the reflection parameter b. Figure 5-25 shows a cross section of the growth rate σ_r and the frequency σ_i. As the coupling coefficient becomes larger, the solution forks into two branches, one extending upward, the other downward. The frequency is zero on the right of the branching point. The instability of the other two solutions of T_0 other than 0 in Eq. (5-65) for the region $a > b$ is sketched in Fig. 5-24(b). The solutions are unstable when the reflection parameter b is relatively large. In short, the results of the instability analysis can be classified into four regions. They are: (1) The region where T_0 is stable. (2) The region where the only unstable

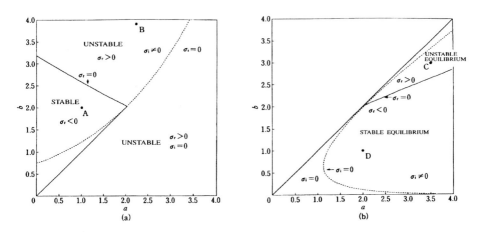

Fig. 5-24. Stability of the delayed oscillator. Linear stability of equilibrium solutions (a) $T_0 = 0$, (b) $T_0 = \pm\sqrt{(a-b)/c}$ (existing in the region $a > b$). $\sigma_r > 0$ ($\sigma_r < 0$) indicates stable (unstable).

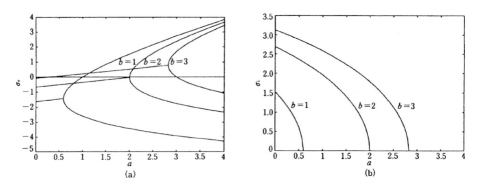

Fig. 5-25. (a) Growth rate (year^{-1}) and (b) frequency (year^{-1}) for coupling coefficients.

equilibrium solution $T_0 = 0$. (3) The region where there are equilibrium solutions of non-zero T_0, which are unstable. (4) The region where there are solutions of non-zero T_0, which are stable.

Battisti and Hirst (1989) estimated representative parameters at the point B in Fig. 5-24: $a = 1/2.2$ years, $b = 1/3.9$ years, $\tau = 180$ days, and $c = 0.07/K^2/y$. They are in the region (2) of linear instability. However, Suarez and Schopf (1988) who first introduced "the delayed oscillator equation" were considering the point C, which is in the region (3).

The solutions of the delayed oscillator equation using a numerical method are shown in Fig. 5-26 for various parameters of (a, b). In the top figure, which corresponds to the point A $(1.0, 2.0)$ in Fig. 5-24(a) in the region (1) where T_0 is the only stable equilibrium solution, the temperature damps to zero and no El Niño takes place. At the point B $(2.2, 3.9)$ in the region (2) where the solution $T_0 = 0$ is unstable, an oscillatory solution appears (the second figure from the top). At the point C $(3.5, 3.0)$ in the region (3) where three equilibrium solutions exist but all of them are unstable, the solution also oscillates (the third figure). At the point D $(2.0, 1.0)$ in the region (4) where two stable equilibrium solutions exist, one of them occurs depending on the initial condition to be selected. Thus, the delayed oscillator equation can reproduce an oscillation with a very long period. It also indicates that stable equilibrium solutions are possible for certain parameters as found in some coupled atrmosphere-ocean models.

5.5.4 Linear eigenvalue problem for a closed ocean basin

Let us think about where the difference of the results of various models comes from. Since the basic stuctures of the models are quite similar, does the difference come from the parameters? To answer the question, it would be best to solve the equation varying parameters in the whole range, which is not easy. The time-evolution model is not appropriate for such a problem. The linearized eigenvalue problem discussed in Section 5.3 would be more effective. From this

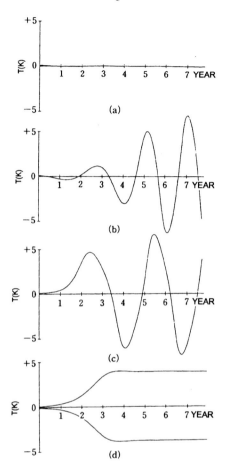

Fig. 5-26. Time evolution of the delayed oscillator equation. Parameters (a, b) are selected as A(1.0, 2.0), B(2.2, 3.9), C(3.5, 3.0), D(2.0, 1.0) at the points in Fig. 5-24.

viewpoint, the instability of the solution to an eigenvalue problem for a closed ocean basin was studied (Hirst, 1988; Wakata and Sarachik, 1991). What kinds of parameters are important then? The candidates will be sought from the thermodynamics equation

$$\frac{\partial T}{\partial t} = -u\overline{T}_x + K_T h - dT \qquad (5\text{-}68)$$

where the parameters are $d \equiv (\overline{W}_s/H_1 + n_d)$ and $K_T \equiv (\overline{W}_s \gamma/H_1)\!<\!AB\cos h^{-2} B\,\overline{h}\!>$. These are introduced from the procedure of linearization, and are expressed by the variable of the mean field such as the entrainment velocity ($\overline{W}_s$) and the depth of the thermocline $\bar{h}$ as discussed in Section 5.2.4.

Hirst calculated for various combinations of K_T and T_x which are spatially uniform but did not succeed in deriving a solution of standing oscillation type. He obtained a solution that propagates to the east using typical parameters for the tropical Pacific (see Section 5.5.1). Wakata and Sarachik (1991) noted spatial non-uniformity of K_T and d, and assumed the thermocline depth of the basic field as

$$\overline{h(x)} = -P\left(\frac{x - x_c}{R}\right) + Q.$$ (5-69)

For the typical values of $P = 50$ m, $Q = 100$ m, $x_c = 10,350$ km and $R = 2,300$ km, the mixed layer depth will be 150 m in the western end of the Pacific, and 50 m in the eastern end. These values are in reasonable agreement with observations. Since the equatorial upwelling is confined to the narrow equatorial belt, the mean upwelling velocity was assumed to have the form

$$\overline{W_s} = \left\langle \overline{W_s} \right\rangle \exp\left(-\alpha y^2\right)$$ (5-70)

where $<W_s>$ is an upwelling velocity on the equator and $\alpha^{-1/2}$ is a meridional e-folding length. From the Ekman boundary-layer therory, the typical values in the tropical Pacific are $<W_s> = 2$ m/day and $\alpha^{-1/2} = 146$ km. The above assumptions result in sharp changes of K_T in the north-south direction, unlike the calculation by Hirst.

For an ocean bounded by walls in the east and west, an eigenvalue problem was solved, by expanding each variable in an even Hermite function series in the north-south direction and replacing the derivatives in the east-west direction with differences. As a result, only one out of many eigenmodes was found to be unstable. This mode had a growth rate of 1/222 days, a period of 910 days (2.5 years). Figure 5-27(a) shows the amplitude structure of the thermocline depth anomaly of the unstable mode. Note that the solution is symmetric about the equator. The values are normalized so that the maximum SST anomaly is 2 K. One peak can be seen at 1.3 units north (4.2°N) in the western Pacific, which corresponds to the peak of the depth of the gravest equatorial symmetric Rossby mode. The other peak can be seen on the equator in the eastern Pacific, which indicates that the Kelvin wave mode has a large amplitude there. The phase of the depth anomaly is shown in Fig. 5-27(b). The phase convention is chosen so that the phase is zero at the coldest state of the ENSO. The change of the phase is prominent in the central Pacific near the equator, showing the eastward propagation of the depth anomaly prior to the warm phase. Wide areas in the western and eastern Pacific have almost unchanged phases. This result suggests that there is an oscillation of a standing wave type where the mixed-layer depth goes up and down as if it were a seesaw. The amplitude of the sea surface

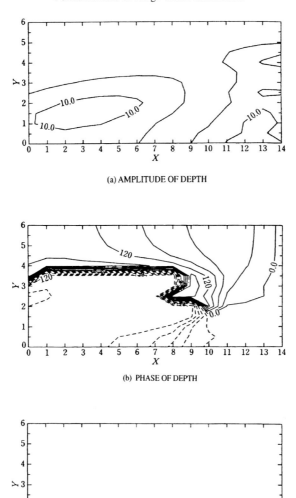

(a) AMPLITUDE OF DEPTH

(b) PHASE OF DEPTH

(c) AMPLITUDE OF TEMPERATURE

Fig. 5-27. An atmosphere–ocean coupling unstable mode in a basin (Wakata and Sarachik, 1991). (a) Amplitude of the thermocline depth anomaly of the unstable mode. The grid interval is 5 m. The X-axis is a longitude coordinate with unit σx (115.1 km). The Y-axis is a latitude coordinate with unit σy (360 km) which is a Rossby deformation length. The origin is the equator, and the pattern is symmetrical on either side of the equator. The values are normalized as the maximum temperature becomes 2 K. (b) Phase of the thermocline depth anomaly of the unstable mode (wave propagates toward phase increasing). The phase is defined as zero when the sea surface temperature anomaly attains the minima. (c) Amplitude of the sea surface temperature. The contour interval is 0.29 K.

temperature anomaly is shown in Fig. 5-27(c). The peak, and indeed most of the amplitude, is seen in the equatorial eastern Pacific.

The strength of the ocean–atmosphere coupling is measured by the product of the wind stress parameter K_S, and the heating parameter K_Q. The nondimensional value of this parameter $K_Q K_S$ was 20 for the standard case of the example mentioned above. We are going to see how the growth rate and the period will vary if $K_Q K_S$ is changed. Figure 5-28 shows the growth rate and the frequency for $K_Q K_S$. With the $K_Q K_S$ increasing, the growth rate of the coupled mode increases and the frequency decreases (the period increases). The growth rate splits into two branches at $K_Q K_S = 32.5$ where the frequency becomes zero. This is more or less similar to the result of the delayed oscillator (Fig. 5-25).

Returning to the main topic, why do we have an eastward-propagating wave or a standing wave, depending on the models? To understand this discrepancy, we need to pay attention to the mixed-layer depth of the basic field and the spatial non-uniformity of the upwelling. The mixed-layer is deep in the west and shallow

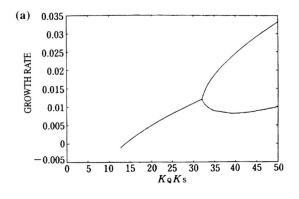

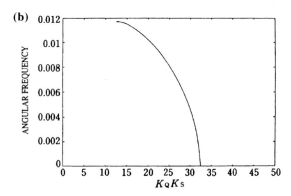

Fig. 5-28. (a) Growth rate (1.0×10^{-2} corresponds to an e-growth rate of 142 days). (b) Frequency (corresponds to a period of 892 days) (Wakata and Sarachik, 1991).

in the east in the tropical Pacific. The difference is given by P in Eq. (5-69). The meridional variation of the upwelling is represented by $\alpha^{-1/2}$. Figure 5-29 shows the phase difference of the depth anomaly between the points $X = 9$ and $X = 13$ (see Fig. 5-27) as a function of thermocline tilt and the meridional extent of upwelling, after solving the eigenvalue problem by varying these parameters. A large positive value means a strong eastward propagation, and a negative value a westward propagation. For a standing wave, the phase should be zero. The anomaly becomes non-propagating when the mean upwelling is tightly confined meridionally, while it propagates rapidly in the case of a wider upwelling extent. This result can explain why only propagating anomalies appeared in the Anderson and McCreary (1985) and Hirst (1986) models, both of which correspond to the lower right corner ($P \to 0$, $\alpha^{-1/2} \to \infty$). On the other hand, the Zebiac and Cane (1987) model corresponds to the upper left corner of the figure ($P = 50$, $\alpha^{-1/2} = 1$). It demonstrates that the propagation characteristics are more sensitive to the meridional extent of the mean upwelling velocity than to the east-west tilt of the mean thermocline depth. This is because the coupling stability of the equatorial Rossby waves that have a maximum outside of the strong upwelling region near the equator plays an important role in distinguishing the characteristics of the models (Wakata and Sarachik, 1992). In the Anderson and McCreary model, the equatorial Rossby waves decay rapidly

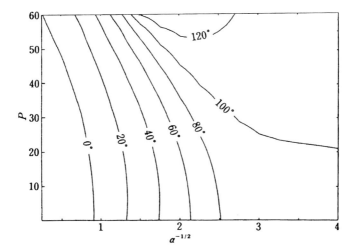

Fig. 5-29. Influence of the meridional width of the upwelling velocity of the mean field on the phase propagation properties of the unstable mode (Wakata and Sarachik, 1991). Phase difference of thermocline depth anomaly between the two points $X = 9, 13$ along the equator in Fig. 5-27. The transversal axis indicates the meridional width of the strong upwelling area, of which unit is 146 km. The ordinates axis indicates the east-west inclination of thermocline. $2P$ corresponds to the east-west difference of thermocline depth. $P = 50$ m is a standard value.

as a result of the ocean–atmosphere interaction, and the growing coupled mode is more like an eastward-propagating Kelvin wave (see also Yamagata and Masumoto, 1989). In the Zebiac and Cane model, however, the equatorial Rossby waves do not decay, and are reflected by the western boundary, creating a delayed oscillation. In order to constitute a standing wave, at least two waves, one eastward-propagating, the other westward-propagating, are necessary. If either decays, no standing wave will be formed.

5.6 WHAT COMES NEXT?

We have so far introduced, using relatively simple models, the results of researches on large-scale air–sea interactions, which have progressed rapidly since the early 1980s. In this section we will give an overview of how these results are being developed and utilized.

The TOGA (Tropical Ocean and Global Atmosphere) program under WCRP (World Climate Research Program), promoted by WMO (World Meteorological Organization), IOC (Intergovernmental Oceanographic Commision) and ICSU (International Council of Scientific Unions), was completed in 1994. The results of the 10-year work include setting up of 60 or so buoys in the tropical Pacific and building a system of monitoring hydrographic and meteorological data in real time via satellites. These data are dynamically interpolated or extrapolated in ocean models (data assimilation) to be used for maps of current and seawater temperature distributions in the tropics, which are now operationally done by the National Meteorological Center (US) and the Japan Meteorological Agency. One of the outstanding results of the meteorological and oceanographical researches in the 1980s is starting such a work-site operation. An attempt has been made to assimilate the data to a coupled model to predict short-term climate change. Figure 5-30 is the result of Neelin *et al.* (1994) showing the effectiveness of the coupled model by Cane *et al.* (1986) for predicting sea surface temperatures. At a glance, the predicted temperatures appear to be no good for the first four or five months, but this is because the coupled model at that time did not use the observed sea surface temperatures as initial values. If observed temperatures had been assimilated initially, the predicted results would have been much better for a year or longer. Hence, all developed countries are now competing to develop assimilation techniques for a full-scale ocean–atmosphere-coupled model using oceanic and atmospheric observations including satellite data. Short-term climate prediction will hopefully be made possible on an operational basis in the near future. Forecasting El Niño will not be especially difficult once a precursor in the ocean–atmosphere system is found, although its prediction beyond spring when a new seasonal thermocline is formed might have some problems (spring barrier). In particular, since the development and decay take a year, it is possible to forecast and take measures beforehand if once some precursor could be captured in the atmosphere–ocean coupled model. In fact, Peru suffered from great damages in agricultural production in the 1982–1983 El Niño, but in 1986–1987, succeeded in increasing its production by taking the precaution of decreasing

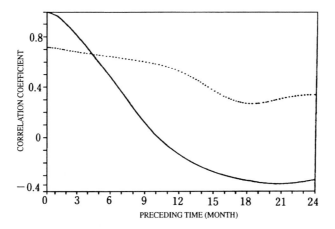

Fig. 5-30. Skill scores of prediction ensembles as a function of lead time for forecasts by the Zebiak and Cane (1987) coupled model (dotted curve), compared with skill obtained by assuming persistence of anomalies (sold curve). The measure is correlation of predicted and observed SST averaged over the region of largest ENSO anomalies (±5° latitude, 150°W to 90°W longitude), during the period from 1972 to 1991. Note that 0-month model score is inferior to observed because SST data are not used in the initialization. Data from S. Zebiak (personal communication); for methodology see Cane et al. (1986).

cotton trees fit for dry and cool weather and increasing rice fit for warm and damp weather before the onset of El Niño.

Oceanic and atmospheric scientists have shifted their interest in how they can find precursors of El Niño. As discussed in the previous sections, if a positive feedback occurs for some reason, it can develop into a proper mode. The delayed oscillator model attributed El Niño to the oceanic Rossby waves accompanying the preceding La Niña and the Kelvin waves created upon reflection at the western boundary. In a real situation, warm water is sometimes accumulated in a broad area extending to mid latitudes (Wyrtki, 1979), in which case intensified easterly winds, rather than free Rossby waves, could be the more plausible cause of the accumulation. On the other hand, westerly bursts accompanying the Maden–Julian disturbance in the atmosphere often take place in winter in the western Pacific, especially in the year preceding an El Niño event (Yamagata and Masumoto, 1989). Many researchers challenge the origin of irregular occurrences of El Niño from the standpoint of relatively small-scale phenomena like the just mentioned short-term interactions of atmospheric disturbances with the ocean.

Since 1976, trade winds have been weaker in general and sea surface temperatures tend to be high in the central and eastern Pacific (Nitta and Yamada, 1989). That is, a condition of weak El Niño has been maintained. This means El Niño may easily be triggered, whereas typical La Niña is unlikely to occur. At the same time the Aleutian low pressure is intensified in the north Pacific, sea surface pressure is low, wind waves are high and the sea surface temperatures are low due to mixing and evaporation. One explanation of this is a climatic regime shift (Trenberth, 1990). Whether this decadal scale variation is simply a modulation of the interannual variations in the tropics or a quite different phenomenon remains an interesting topic of study. We know empirically that winter in Japan tends to be warmer during El Niño. It is believed that this is because the temperature of seawater is low in the western tropical Pacific then, being not very much different from that of the continent, and the cold surges are weakened (Hanawa *et al.*, 1989). In the past decade or so, the Aleutian Low has been strong, shifted to the south, and the cold winds blowing into it often covers the whole of Japan. This may be the reason why the empirical law about the relationship between the Japanese winter and El Niño does not hold any longer (Yamagata and Masumoto, 1992). Research interest is toward such longer time scale phenomena as global warming and whether it is separable from interdecadal climate variability. In any event, research on the monsoon, or any atmospheric and oceanic research including the Indian Ocean and the Atlantic, namely, the air–sea–land interaction studies will be more and more important in the future.

REFERENCES

Anderson, D. L. T. and J. P. McCreary (1985): *J. Atmos. Sci.*, **42**, 615–628.
Battisti, D. S. (1988): *J. Atmos. Sci.*, **45**, 2889–2919.
Battisti, D. S. and A. C. Hirst (1989): *J. Atmos. Sci.*, **46**, 1687–1712.
Bjerknes, J. (1969): *Mon. Wea. Rev.*, **97**, 163–173.
Cane, M. A. and E. S. Sarachik (1977): *J. Mar. Res.*, **35**, 395–432.
Cane, M. A., S. C. Dolan and S. E. Zebiak (1986): *Nature* (London), **321**, 827–832.
Gill, A. E. (1980): *Quart. J. Roy. Meteor. Soc.*, **106**, 447–469.
Gill, A. E. (1985): Elements of coupled ocean–atmosphere models for the tropics. In *Coupled Ocean–Atmosphere Models*, Elsevier, pp. 303–327.
Gill, A. E. and E. M. Rasmusson (1983): *Nature*, **306**, 229–234.
Hanawa, K., Y. Yoshikawa and T. Watanabe (1989): *J. Meteor. Soc. Japan*, **67**, 385–400.
Hirst, A. C. (1986): *J. Atmos. Sci.*, **43**, 606–630.
Hirst, A. C. (1988): *J. Atmos. Sci.*, **45**, 830–852.
Holton, J. (1975): *Meteorol. Monogr.*, **15**, 216.
Lighthill, M. J. (1966): *Phil. Trans. R. Soc. Lond.*, **A265**, 45–93.
Masumoto, Y. and T. Yamagata (1991): *J. Meteor. Soc. Japan*, **69**, 197–207.
Matsuno, T. (1966): *J. Meteor. Soc. Japan*, **44**, 25–43.
McCreary, J. (1981): *Philos. Trans. R. Soc. London*, **298**, 603–635.
Moore, D. W. and S. G. H. Philander (1977): *The Sea*, **6**, 319–361.
Neelin, J. D., M. Latif and F. F. Jin (1994): *Annu. Rev. Fluid Mech.*, **26**, 617–659.
Nitta, T. and S. Yamada (1989): *J. Meteor. Soc. Japan*, **21**, 1386–1398.
Philander, S. G. H. (1990): *El Niño and La Niña, and the Southern Oscillation*, Academic Press, Inc., 293 pp.
Philander, S. G. H., T. Yamagata and R. C. Pacanowski (1983): *J. Atmos. Sci.*, **41**, 604–613.
Rennick, M. A. (1983): *Tropical Ocean–Atmophere Newsletter*, **15**, 2–4.
Schopf, P. S. and M. J. Suarez (1988): *J. Atmos. Sci.*, **45**, 549–566.
Suarez, M. J. and P. S. Schopf (1988): *J. Atmos. Sci.*, **45**, 3283–3287.
Trenberth, K. E. (1990): *Bull. Amer. Meteor. Soc.*, **71**, 988–993.
Wakata, Y. and E. S. Sarachik (1991): *J. Atmos. Sci.*, **48**, 2060–2077.

Wakata, Y. and E. S. Sarachik (1992): *J. Meteor. Soc. Japan*, **70**, 843–854.

Wyrtki, K. (1979): *J. Phys. Oceanogr.*, **9**, 1223–1231.

Wyrtki, K. and B. Kilonsky (1984): *J. Phys. Oceanogr.*, **14**, 242–254.

Yamagata, T. (1985): Stability of a simple air–sea coupled model in the tropics. In *Coupled Ocean Atmosphere Models*, Elsevier, pp. 637–658.

Yamagata, T. and Y. Masumoto (1989): *Philos. Trans. Soc. London*, **329**, 225–236.

Yamagata, T. and Y. Masumoto (1992): *J. Meteor. Soc. Japan*, **70**, 167–175.

Yamagata, T. and T. Matsuura (1993): *Kagaku*, **63**, 673–677 (in Japanese).

Yamagata, T. and S. G. H. Philander (1985): *J. Oceanogr. Soc. Japan*, **41**, 345–357.

Yoshida, K. (1959): *J. Oceanogr. Soc. Japan*, **15**, 154–170.

Zebiak, S. E. and M. A. Cane (1987): *Mon. Weather Rev.*, **115**, 2262–2278.

Chapter 6

Numerical Modeling for Large-Scale Ocean–Atmosphere Interactions

Masahiro ENDOH, Yoshiteru KITAMURA,
Hiroshi ISHIZAKI and Tatsuo MOTOI

Ocean–Atmosphere Interactions, Ed. Y. Toba, pp. 195–237.
© by TERRAPUB / Kluwer, 2003.

Numerical Modeling for Large-Scale Ocean–Atmosphere Interactions

Masahiro ENDOH[1], Yoshiteru KITAMURA[2],
Hiroshi ISHIZAKI[3] and Tatsuo MOTOI[4]

I was like a boy playing on the sea-shore, and diverting myself now and then finding a smoother pebble or a prettier shell than ordinary, whilst the great ocean of truth lay all undiscovered before me.

—*Isaac Newton*

A good model used for research on ocean–atmosphere interactions should contain physical processes in as simplified forms as possible. An ideal model that would provide the best insight may be expressed in analytical formulae. However, existing air–sea interactions include complex nonlinear physical processes on both spatial and temporal scales, such as heat and momentum transports, convections and unstable eddies. When enough observations are unavailable, it is rather difficult to foresee what kinds of physical processes are involved. Accordingly, a numerical model was introduced to describe the complex nonlinear processes in numerical digits (or data) and a practical method of perceiving nature's mechanism by analyzing such a model has been made possible as high-speed computers have been developed.

In this chapter, methods of numerical modeling frequently used for large-scale oceanic and atmospheric circulations is generally described. Then, researches using ocean models and coupled models on various spatial and temporal scales will be introduced. Aim of this chapter is to understand actual methods of model calculations.

6.1 TOOLS FOR MODELING

6.1.1 Modeling and numerical models

Numerical calculation in earlier times was a means for obtaining mathematical solution of physical processes in a system under consideration. For instance,

---

[1] Sections 6.1 and 6.5
[2] Section 6.2
[3] Section 6.3
[4] Section 6.4

the propagation of oceanic Rossby waves was studied by solving the following vorticity equation with the given external wind force (on the right-hand side).

$$\frac{\partial \nabla^2 \psi}{\partial t} + \beta \frac{\partial \psi}{\partial x} = \frac{\text{curl} \tau}{\rho h}. \tag{6-1}$$

However, it was recognized that the same vorticity equation, which also governs global scale circulations in the atmosphere, was usable for weather forecasting if the solution was obtained starting from an appropriate initial condition, and this led to the concept of numerical modeling. As computers and observation techniques progressed, forecasting of weather and climate was made possible. At the same time, the analysis of numerical values (data set) predicted by a model at spatially arranged grid-points, obtained status as research on natural phenomena. That is, modeling is now a means of studying mechanisms of variations in nature by regarding the data integrated from a physical equation with realistic conditions as if they were observed values. In practice, the vorticity equation alone cannot deal with complex boundary conditions or thermodynamic conditions, so that fundamental equations of hydrodynamics and thermodynamics should also be included, and the set of simultaneous partial differential equations are solved after being converted to corresponding finite difference equations.

6.1.2 Preparation for numerical calculations

a) Types of partial differential equations

In many cases of oceanographic and meteorological problems, partial differential equations describing time evolution are often to be solved. The vorticity equation (6-1) is one example. Other examples are the usual linear wave equation

$$\frac{\partial^2 u}{\partial^2 t} + c^2 \frac{\partial^2 u}{\partial^2 x} = 0, \tag{6-2}$$

and the advection equation given by

$$\frac{\partial u}{\partial t} + c \frac{\partial u}{\partial x} = 0. \tag{6-3}$$

This equation is similar in form to a linearized version of the first two terms in the equation of motion, the thermodynamic or vorticity equations, etc. It describes a wave propagating in the x-direction at speed c without change in form. The wave equation (6-2) is generalization of Eq. (6-3) and this kind of second-order linear partial differential equation is called hyperbolic, in which two families of real characteristics exist. Note that characteristics are lines across which derivatives of the dependent variables may be discontinuous and infinitesimal disturbances or signals can propagate along characteristics. The solutions to Eq. (6-2) are wave-like in character. Its characteristics are given by the lines of $\alpha = x - ct =$ constant and $\beta = x + ct =$ constant. Signals propagate along these lines in space and time starting from spatially distributed initial values.

On the other hand, the Poisson equation given by

$$\nabla^2 \psi = -\rho \qquad (6\text{-}4)$$

is called elliptic, for which no real characteristics exist. Elliptic equations usually describe the steady-state limit of problems where the time-dependent problem is described by hyperbolic or parabolic equations. The most common parabolic equation is the diffusion equation of the Fickian type given by

$$\frac{\partial u}{\partial t} = \kappa \frac{\partial^2 u}{\partial^2 x}. \qquad (6\text{-}5)$$

For this equation, only one family of characteristics exists: they are $t =$ constant. Parabolic equations often arise from hyperbolic equations in the limit as the propagation velocity becomes infinite, a good approximation in many cases where the velocity of wave propagation is much greater than other velocities present in the problem. In this limit, it follows that a disturbance at any point at a particular time is felt instantaneously at all other points of the spatial domain.

The above discussion was based on purely mathematical classification of the second-order partial differential equations, but from a viewpoint of mathematical physics, they can better be grouped into two major categories: initial value problems and boundary value problems. They are important concepts when we consider how to solve them numerically.

In order to solve oceanographic and meteorological problems, a complicated system of simultaneous equations combining the above three types should be used. However, in the subsequent chapters, the simplest advection equation, Eq. (6-3) will be mostly considered, unless mentioned otherwise. We are focusing more on a practical introductory method than a systematic description of numerical methods for complex equations. It may be worthwhile to mention that Eq. (6-3), a first order partial differential equation, has real characteristics and, because of this, behaves in many ways like a hyperbolic equation of the second order.

b) Grid point method and finite difference approximation
Here we will describe the technique of using the grid point method, and the problems associated with it, using the grid of uniform computation points fixed in space and time. In real applications, it is sometimes more desirable to adopt

non-uniformly spaced grid points in regions where the solution is changing rapidly, such as in the equatorial or the western boundary region in the ocean. In case where the spherical coordinate system is used, independent variables (longitude and latitude) are uniformly spaced, whereas grid points in the corresponding physical space are non-uniformly distanced. In the isopycnal coordinate system, grid point intervals vary with time.

With the grid point method, the most common way of solving differential equations is to find approximate expressions for derivatives appearing in the equations. These approximate expressions are defined using only values of the dependent variables at the grid points, and at discrete time intervals. Thus they are formed using differences of dependent variables over finite space and time intervals. For this reason, this approach is referred to as the finite difference method. Let us assume that a function $u(x, t)$ is smooth with no singular points. Let Δx and Δt be uniform distance between two grid points in space and time, respectively, j be the sequential number for each grid point. If u_j is the value for the variable at the j-th coordinate grid (x or t), then a forward difference at the point j is defined as $\Delta u_j \equiv u_{j+1} - u_j$ and a centered difference as $\delta u_j \equiv u_{j+1/2} - u_{j-1/2}$. For instance the partial derivative at the point j, $(\partial u / \partial x)_j$, is written as $\Delta u_j / \Delta x$ in the forward difference approximation.

We now estimate the truncation error for the most often used forward difference approximation. In general, by expanding $u(x + \Delta x)$ in a Taylor series about the point j, we obtain

$$\frac{u_{j+1} - u_j}{\Delta x} = \left(\frac{\partial u}{\partial x}\right)_j + \frac{1}{2!}\left(\frac{\partial^2 u}{\partial x^2}\right)_j \Delta x + \frac{1}{3!}\left(\frac{\partial^3 u}{\partial x^3}\right)_j (\Delta x)^2 + \cdots. \qquad (6\text{-}5')$$

Then the error due to the forward difference approximation will be

$$\varepsilon \equiv \frac{u_{j+1} - u_j}{\Delta x} - \left(\frac{\partial u}{\partial x}\right)_j = \frac{1}{2!}\left(\frac{\partial^2 u}{\partial x^2}\right)_j \Delta x + \frac{1}{3!}\left(\frac{\partial^3 u}{\partial x^3}\right)_j (\Delta x)^2 = O(\Delta x). \qquad (6\text{-}5'')$$

We can say that the accuracy of the error is of the order Δx in this case. Similarly, a forward difference with respect to time $(u^{n+1} - u^n)/\Delta t$ has an accuracy of Δt. On the other hand, a centered difference will be

$$\varepsilon \equiv \frac{u_{j+1} - u_{j-1}}{2\Delta x} - \left(\frac{\partial u}{\partial x}\right)_j = \frac{1}{3!}\left(\frac{\partial^3 u}{\partial x^3}\right)_j (\Delta x^2) = O(\Delta x^2). \qquad (6\text{-}5''')$$

In this case the error is Δx^2: the centered difference is a second-order approximation and has better accuracy than the forward difference. For this reason the centered difference is more often used.

6.1.3 Fundamentals for numerical methods

a) Convergence and stability of solutions

Let us take the partial differential equation, Eq. (6-3) and consider the forward difference with respect to time and the central difference with respect to space.

$$\frac{u_j^{n+1} - u_j^n}{\Delta t} + c\frac{u_{j+1}^n - u_{j-1}^n}{2\Delta x} = 0 . \qquad (6\text{-}6)$$

This equation converges to Eq. (6-3) as Δx, $\Delta t \to 0$: in other words, it has consistency. When integration is performed using Eq. (6-6), the calculation of $u^{n+1}{}_j$ requires variables one step before, namely, $(u^n{}_j, u^n{}_{j+1}, u^n{}_{j-1})$. If we further go back some time-steps, the domain of dependence in the $(\Delta x, \Delta t)$-grid point box will extend to the past radially. Difference calculation is always done in this domain of dependence, so that the gradient $1/c$ of the dependence line $(t = x/c)$ for the true solution $(u = f(x - ct))$ must satisfy the following condition in order for the solution of Eq. (6-6) to converge to the solution of Eq. (6-3).

$$\frac{1}{c} \geq \frac{\Delta t}{\Delta x} : \quad \text{(Conversion condition)} . \qquad (6\text{-}7)$$

However, it is not guaranteed that the solution of the difference equation, Eq. (6-6) is neutrally stable as time elapses $(n \to \infty)$ like the solution of the original Eq. (6-3). Von Neumann, studying the stability of partial differential equations, showed that the amplitude ratio $\lambda = U^{n+1}/U^n$ for $u = Re(U(t)e^{ikx})$ must satisfy $|\lambda| \leq 1 + O(\Delta t)$ as a necessary condition for the solution to be stable. If we adopt the central difference in time and space for Eq. (6-3), a simple algebraic calculation after substituting $u^n{}_j = Re(U^{(n)}e^{ikj\Delta x})$ can show that the von Neumann stability condition in this case will become

$$c\frac{\Delta t}{\Delta x} < 1 \qquad (6\text{-}8)$$

which is commonly referred to as the CFL (Courant–Friedrichs–Levy) condition for computational stability. It should be noted that Eq. (6-8) is a necessary, but not sufficient, condition.

b) Time-integration schemes

Schemes used for integrating the time derivative terms are essentially the same as those for spatial derivative terms, but there are some differences. The former are relatively simple, usually of the second order and sometimes even only of the first order of accuracy. One reason for not requiring a scheme of high

accuracy for approximations to the time derivative terms is that the error of the
numerical solution in most oceanic and atmospheric models is brought about
more by insufficient information about the initial conditions than by the
inadequacy of the scheme itself. Another reason is that, in order to meet a stability
requirement Eq. (6-8), it is usually necessary to choose a time step significantly
smaller than that required for adequate accuracy. With the time step usually
chosen, other errors, for example in the spatial terms, are much greater.

Characteristics of various schemes are summarized below.

(i) Two-level schemes

For the equation $\partial u/\partial t = f(u, t)$, we want to construct a scheme for
computation of an approximate value u at time $(n + 1)\Delta t$, assuming the values of
previous time $n\Delta t$, $(n - 1)\Delta t$, ... are all known, where Δt is the length of a segment
in the time axis. If we write formally,

$$u^{(n+1)} = u^{(n)} + \Delta t\left(\alpha f^{(n)} + \beta f^{(n+1)}\right) \qquad (6\text{-}9)$$

where $\alpha + \beta = 1$ and $f^{(n)} = f(u^{(n)}, n\Delta t)$, then the schemes can be classified
according to the values of α and β.

(1) Forward (or Euler) scheme ($\alpha = 1$, $\beta = 0$): In this case Eq. (6-9) will be
simplified to $u^{(n+1)} = u^{(n)} + \Delta t f^{(n)}$. The truncation error of this scheme is
the first order of Δt.

(2) Backward scheme ($\alpha = 0$, $\beta = 1$): The formula Eq. (6-9) will become
$u^{(n+1)} = u^{(n)} + \Delta t f^{(n+1)}$. Since a value of f depends on $u^{(n+1)}$ in this case,
this scheme is implicit: it requires solving a set of simultaneous
equations, with one equation for each of the grid points of the computation
region. (Note that the forward scheme is explicit.)

(3) Trapezoidal scheme ($\alpha = 0.5$, $\beta = 0.5$): This is an intermediate between
the forward and backward schemes approximating f by an average of the
values at the beginning and the end of the time interval:

$$u^{(n+1)} = u^{(n)} + \Delta t(f^{(n)} + f^{(n+1)})/2 \ .$$

(4) Matsuno (or Euler-backward) scheme: With this scheme a step is made
first using the forward scheme; the value of u obtained for time step
$(n + 1)$ is then used for an approximation to $f^{(n+1)}$ and this approximate
value $f^{(n+1)*}$ is used to make a backward step. Thus,

$$\begin{aligned}
u^{(n+1)*} &= u^{(n)} + \Delta t \cdot f^{(n)} \ , \\
u^{(n+1)} &= u^{(n)} + \Delta t \cdot f^{(n+1)*} \ .
\end{aligned} \qquad (6\text{-}10)$$

This method can selectively attenuate high frequency components.

The stability of the schemes described above can be examined by considering the case $f = i\omega u$. This is equivalent to considering the oscillatory solution with $\omega = kc$ in the advection equation, Eq. (6-3). Therefore, the exact analytical solution will be $u(t) = u(0)e^{i\omega t}$, and the solution for its difference equation $u(n\Delta t) = e^{in\omega\Delta t}$. By substituting $f^n = i\omega u^n$ in Eq. (6-9), we obtain $\lambda = u(n + 1)/u(n) = (1 + i\alpha p)/(1 - i\beta p)$ where $p = \omega\Delta t$. Here λ is called the amplification rate. A scheme is unstable (amplifying) if $|\lambda| > 1$, neutral if $|\lambda| = 1$, and stable (damping or dissipative) if $|\lambda| < 1$.

For the forward scheme $|\lambda| = (1 + p^2)^{1/2}$ so that it is unstable regardless of the value of p or Δt.

For the backward scheme $|\lambda| = 1/(1 + p^2)^{1/2}$ so that it is always stable.

For the trapezoidal scheme $|\lambda| = 1$, so that it is neutral.

For the Matsuno scheme $|\lambda| = (1 - p^2(1 - p^2))^{1/2}$. If $|p|<1$ ($\Delta t < 1/\omega$), then $|\lambda| < 1$. Thus this scheme is stable and it dampens high-frequency components of ω close to $1/\Delta t$.

(ii)   Three-level schemes

One can construct schemes taking advantage of additional information from time step $(n - 1)$ except at the first step. These are three-level schemes. The simplest and most widely used is the leapfrog scheme

$$u^{(n+1)} = u^{(n-1)} + 2\Delta t f^{(n)} .$$

(6-11)

This has the second order of accuracy. The stability analysis leads to $\lambda = \pm(1 - p^2)^{1/2} + ip$. Thus when $|p| < 1$ ($\Delta t < 1/\omega$), which usually holds, $|\lambda|$ is equal to 1. Therefore, the scheme is ideal in the sense that it is unconditionally neutral. However, there are two solutions of the form $u^{(n+1)} = \lambda u^{(n)}$. This necessarily follows from the fact that we are considering a three-level scheme. As $p \to 0$, λ approaches either 1 or -1. The former corresponds to the ture solution (physical mode), whereas the latter is a computational mode. We know empirically that using a forward scheme every few dozen time steps of the leapfrog scheme can suppress the computational mode.

c) Boundary value problem

In many oceanic and atmospheric models, we often obtain the pressure field from the divergence field, or the stream function from the vorticity field. In such cases the equations are called diagnostic, a term used to describe relationships between physical variables in the absence of time derivatives. Mathematically this requires solving elliptic equations over a region, with conditions provided on the boundary enclosing the region. The best-known elliptic equation is the Poisson type given by Eq. (6-4).

The most common method of solution used by oceanographers and meteorologists is the relaxation method: a successive approximation in which an initial guess of the solution is made and then progressively improved until an acceptable level of accuracy is reached. This method is very flexible and can be used under a wide range of conditions, including irregular boundaries, interior boundary points, three dimensions, etc., compared with direct methods which involve solving a multi-dimensional inverse matrix. To illustrate the relaxation method, consider the following finite difference form of $\nabla^2 u$ at a grid point (i, j).

$$\nabla^2 u_{ij}^n = \frac{u_{i+1,j}^n + u_{i-1,j}^n - 2^n u_{i,j}}{\Delta x^2} + \frac{u_{i,j+1}^n + u_{i,j-1}^n - 2u_{i,j}^n}{\Delta y^2} \qquad (6\text{-}12)$$

where u_{ij}^n is the value of u at the n-th estimate. In the relaxation method the residuals for the n-th estimate $R_{ij}^n = \nabla^2 u_{ij}^n + \rho_{ij}$ are reduced to some acceptably small value (convergence condition) although the exact solution with the $R_{ij}^n = 0$ everywhere will not be reached. Given the n-th estimate, an improved value at the $(n + 1)$th estimate, which will temporarily reduce the residual R_{ij}^n to zero, may be obtained. However, since the residual at any point is dependent on the four surrounding points, it is evident that when the residual at a neighboring point, say $(i + 1, j)$, is reduced to zero in similar fashion, the residual at point (i, j) will depart from zero again. So the successive procedure must be repeated until every grid point satisfies the convergence condition. This procedure, which is called simultaneous relaxation (or more conventionally the Richardson method), always converges, although it may do so slowly.

Some improvements to simultaneous relaxation will yield more rapid convergence. Using new values at preceding points as they are obtained, say, $u^{n+1}_{i-1,j}$ and $u^{n+1}_{i,j-1}$ instead of $u^n_{i-1,j}$ and $u^n_{i,j-1}$ when calculating the residual gives faster convergence and is known as sequential relaxation (or the Liebmann method).

Also it is quickly seen that if the residuals have the same sign over at least several adjacent grid points, it pays to overrelax, that is, to add a larger correction so that the residual becomes $-vR_{ij}^n$, rather than zero (v being the overrelaxation coefficient: $0 < v < 1$). That is, we use the modified value of $u^{n+1}_{ij} = u^n_{ij} + (1 + v)R_{ij}^n/2(1/\Delta x^2 + 1/\Delta y^2)$. This procedure is called overrelaxation (or the accelerated Liebmann method). The optimal value of the coefficient v depends on $\rho(x, y)$, coefficients of the partial derivatives in the equation, the shape of the boundary, etc. It may be very sensitive and small changes can, on occasion, radically change the rate of convergence.

6.1.4 *Expression for physical processes whose spatial scales are smaller than grid intervals*

a) *Diffusion equation and computational stability*

Both oceanic and atmospheric phenomena are distributed over a wide energy spectrum range, and their components interact with one another. For this reason, it is important to express quantitatively the effects of energy, vorticity and momentum transfer to and from smaller-scale motions which cannot be caught by grid points. (For instance, the mixing effect of mesoscale eddies in a general ocean circulation model with coarse grid points.) In general, the mixing effect is formulated based on the mixing length theory of small isotropic turbulence in an analogy to molecular viscosity and diffusion: $\partial u/\partial t = \partial x[F(x)]$, $F(x) = \sigma(x)\partial u/\partial x$. $F(x)$ is called turbulent flux, and $\sigma(x)$ the viscosity or diffusion coefficient.

When the viscosity coefficient $\sigma(x)$ is constant, the above formula is simplified to: $\partial u/\partial t = \sigma\partial^2 u/\partial x^2$. The well known Ekman viscosity and Rayleigh damping can be expressed as

$$\frac{\partial u}{\partial t} = -\kappa u, \quad \kappa = \sigma k^2 . \tag{6-13}$$

Now we will examine the stability of this equation.

(i)   Two-level schemes:

$$\frac{u^{n+1} - u^n}{\Delta t} = -\kappa\left(\alpha u^n + \beta u^{n+1}\right), \quad \alpha + \beta = 1 . \tag{6-13'}$$

In this case the amplification rate is $\lambda = |u^{n+1}/u^n|$, which should be equal to or less than 1 for the scheme to be stable. That is, the forward scheme ($\alpha = 1$, $\beta = 0$) is stable if $\kappa\Delta t < 1$, or $\sigma\Delta t/\Delta x^2 < 1$. Both the backward scheme ($\alpha = 0$, $\beta = 1$) and the trapezoidal scheme ($\alpha = \beta = 1/2$) are unconditionally stable.

(ii)   Three-level schemes:

In case $(u^{n+1} - u^{n-1})/2\Delta t = -\kappa u^n$, one of the two solutions of λ is less than -1, that is, the scheme is invariably unstable. In this regard careful attention should be paid when the leapfrog method is used.

b) Expression for diffusion

How to express the diffusion effect is not directly linked to the method of numerical computations. However, when we make numerical modeling for atmosphere and ocean, we must understand the significance of smaller-scale mixing and select one appropriate formulation. For this purpose the conceptual classification of eddy viscosity and eddy diffusivity will be introduced here.

(i)   Geostrophic (mesoscale) eddies and isopycnal mixing

In the past when calculating large-scale oceanic and atmospheric circulations, grid intervals of larger than a few hundred kilometers with vertical resolution of up to a few hundred meters were normally used assuming isotropy in horizontal directions. The horizontal diffusion coefficient was assumed to have a value between 10^2 m²/sec and 10^5 m²/sec, by analogy to molecular diffusion of chemical substances. However, since the discovery of mesoscale eddies, diffusive effects of geostrophic eddies, which correspond to cyclones in the atmosphere, have been taken into account. They are, in one word, the mixing by eddies on a large-scale isopycnal surface. Since the isopycnal surface is not horizontal, the effect of diffusion spreads systematically in the vertical direction, too. Some concrete formulas were proposed and discussed (Danabasoglu *et al.*, 1994).

(ii) Forced vertical mixing (Mechanical mixing)

In the vertical direction, turbulence by small-scale winds and internal wave breaking are believed to play a dominant role in the ocean. As the diffusion coefficient, an empirically determined value of $(0.5–100) \times 10^{-4}\,\mathrm{m^2\,s^{-1}}$ has been used. On the other hand, the eddy viscosity coefficient $\sigma(z)$ which takes into account the effect of density stratification of the field and the vertical gradient of the current (Richardson number) is proposed.

$$\sigma(z) = \frac{v_0}{(1 + \alpha Ri)^2} + v_b,$$

$$Ri = -\frac{\dfrac{g}{\rho_0} \cdot \dfrac{\partial \rho}{\partial z}}{\left(\dfrac{\partial U}{\partial z}\right)^2} \quad : \text{Richardson number for the field}.$$

(6-13″)

In recent general circulation models, nonlinear turbulence is considered (closure model), and the coefficient $\sigma(z)$ is directly calculated together with the turbulent energy in the field (e.g., see Mellor and Yamada, 1982).

(iii) Density convection and convective adjustment

Vertical convection in an unstable stratification is a very important process in forming thermohaline circulation, but its horizontal scale is about the same as or smaller than the vertical scale so that grid intervals of larger than 10 km cannot incorporate it. Thus a conventional method called convective adjustment is normally used. This is a very simple method in which, when an instability of density occurs between two vertical grid points, the fluid there is mixed isentropically (potential densities of the two points are made equal). The results of many studies show that this method is quite effective for simulating large-scale density fields. But no quantitative evidence has been found yet and its validity is being extensively investigated by using models that express the convective adjustment directly by grids.

6.1.5 Before starting calculation

Now, are we ready for numerical modeling? We seem to have arrived only at the foot of a mountain whose top is far ahead. In subsequent chapters we are going to introduce most recent researches on modeling, omitting the middle part of the mountain which had better be studied thoroughly. Since lack of space does not allow full description of this difficult part, we intend to summarize its essence in the following paragraph.

To start with, you postulate a problem mathematically. It is necessary to parameterize physical quantities. Grid interval is related with viscosity and diffusion coefficients. Several grid points are normally needed within the spatial scale under consideration. Here you have to find a point of compromise between physics and computational stability, by an iterative procedure. Then you draw a flow chart to make a trial run of the program. To detect an unexpected error, which is often contained in a new model, a foolproof test (e.g., zero output for zero forcing) should be devised. At this step necessary computer resources can be estimated so that the program is rewritten to be more efficient. If the resources are insufficient, which is often the case, go back to the first step at the beginning of this paragraph. Then another iterative process will be started until it converges.

6.2 MODELING OF OCEANIC CIRCULATIONS IN MID AND LOW LATITUDES

In this section, we introduce some concrete examples of numerical modeling by illustrating researches on oceanic circulations in mid and low latitudes. The contents and background of the researches are described only roughly, but they are all monumental works in the history of ocean dynamics. Readers are highly recommended to refer to the original publications.

6.2.1 Modeling of western boundary currents

a) Wind-driven circulation and thermohaline circulation

Strong currents such as the Kuroshio and the Gulf Stream exist in the western end of an ocean, and they have extensive influence over climate, transportation and economy in the region. These western boundary currents are also in charge of heat transport from the low to high latitudes, and thus they play an important role in ocean–atmosphere interactions. The wind-driven circulation theory by Stommel (1948) was the first to provide the dynamical explanation about the formation of western boundary currents. He regarded the western boundary current as a compensation current for the internal region where the vorticity of winds is in balance with the planetary β-effect, and obtained a boundary layer where the dissipation of vorticity is predominant due to friction. On the other hand, Robinson and Welander (1963) proposed the thermohaline circulation theory in which the current is driven by temperature difference between north and south. In this case, unlike in the wind-driven circulation, no total transport is generated, but baroclinic western boundary current is formed, that is, the upper layers generate flows in opposite directions in the boundary. These two analytical theories were derived on certain assumptions and the combined effect of both wind and heat was not discussed at all. Therefore, it was not possible to determine which of the two theories was more dominantly applicable to the real ocean circulation, or what kind of balance was holding in the real western boundary current. Thus in order to bring a conclusion to these unsettled problems, studies on general oceanic circulations using numerical modeling were started as computer technology progressed. The most representative work in the early days by Bryan and Cox (1967, hereafter referred to as BC) will be described below.

b) Numerical model of oceanic general circulation

In BC, an ocean basin of uniform depth is considered. It is bounded laterally by two meridians 85° wide, and two latitude circles between 10°N and 70°N, as a model of the Atlantic Ocean. Vertically there are six layers. Temperature and wind stress are specified as functions of latitude at the upper surface. Solutions are obtained by direct numerical integration from a still initial state with uniform density stratification to a steady state. Whether or not they reached the steady state is judged by looking at temporal variations of spatial averages of sea-water temperature, salinity, kinetic energy and available potential energy. In a model containing the western boundary current, as in this case, grid intervals must be designed in such a way that at least one grid point is located in the western boundary region to avoid computational instability. Here as a western boundary, the viscous boundary layer with the width of (viscosity coefficient/planetary β)$^{1/3}$ is often adopted. For instance in case the horizontal eddy viscosity coefficient is 5×10^7 cm^2/sec and the planetary β is 2×10^{-13}/sec, the scale of the boundary layer will be about 100 km. In other words, if the grid interval is 500 km, the viscosity coefficient must have an order of 10^7 cm^2/sec. In BC, the east-west grids are 5° apart in most areas, but the spacing is reduced to 1° in the western boundary region.

The analysis by BC indicates that their system depends on five basic parameters and they conducted eight numerical experiments to investigate the geophysically significant range of these parameters. Any numerical model will easily allow such parametrical experiment, as long as computer time (resource) is available. However, selecting parameters by physical insight is more desirable than varying them randomly. The dimensionless parameters BC obtained physically correspond to: ratio (γ) of the wind stress to the thermohaline effect, ratio of lateral diffusion to inertia, ratio of horizontal viscosity to inertia, ratio of inertia to the Coriolis term, and the effect of vertical diffusion.

Figure 6-1 depicts the result of comparing the thermohaline circulation with the wind-driven circulation, depending on the values of γ. The upper panel shows streamline pattern in the surface layer, and the lower panel illustrates north-south sections of temperature bisecting the basin. It is noted that in the purely thermal case ($\gamma = 1$), the current in the lower layer flows in the opposite direction to that in the upper layer. The streamline patterns show how the single large anticyclonic gyre splits into a subtropical and subarctic gyre as the effect of wind increases. The thermal and wind effects work in the same direction in the subtropical gyre, but tend to oppose each other in the subarctic gyre region, thus making the latter relatively weak. In the case where the wind effect is strongest, the subtropical gyre is nearly symmetric with respect to an east-west axis along latitude 30°N.

In the subtropical gyre the isothermal surfaces show a convex shape downward (lower panel in the figure). Unlike the pure thermal case, the velocity in the southward moving branch of the subtropical gyre increases upward with a pronounced maximum at the surface. Both these tendencies bring the patterns of the solutions into much closer agreement with observations.

6.2.2 Modeling of ventilation

a) Theory of ventilation

Oceans are often considered to consist of two layers above and below the thermocline. Wind-driven circulation does not affect the lower layer very much because of the presence of the thermocline, and thermohaline circulation contributes more to the lower layer than to the upper layer. Then which of the two circulations is greater near the thermocline? As is evident from Fig. 6-1, part of isopycnal lines comprising the thermocline in the tropics, outcrops at the surface in high latitudes due to cooling and mixing. How is the movement of the water determined in this case?

An answer to the above questions is given by the ventilation theory (Luyten

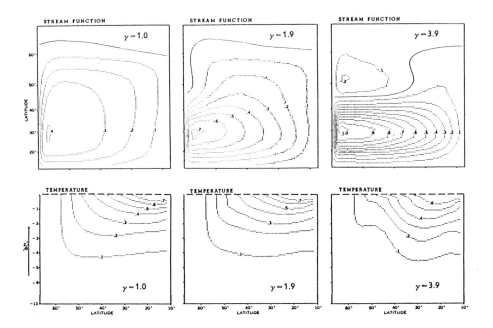

Fig. 6-1. Horizontal patterns of the upper layer stream lines (top), and north-south transects of the temperature at the middle of the basin (bottom) for different γs. γ is a parameter associated with relative intensities of wind stress and sea surface density difference. $\gamma = 1$, 1.9, and 3.9 correspond to density difference plus no wind stress, weak wind stress, and strong wind stress, respectively.

et al., 1983), which explains the structure and mechanism of isopycnal surfaces that outcrop in high latitudes. At the sea surface in the northern part of the subtropical gyre, fluid particles with a certain density and a certain potential vorticity supplied by wind stress and heat flux are subducted into the subsurface due to Ekman convergence (which creates downward flows). After that, since the particles are not directly affected by wind, their motion is regulated by the conservation of potential vorticity. The theoretical solution obtained under this concept shows that a ventilated region where particles sink into the central part of the subtropical gyre is formed. A series of numerical model studies were conducted to test this theory. In the following section, we are going to introduce the result by Cox (1985), which contains the effect of mesoscale eddies.

b) Ventilation numerical model by Cox

Cox's model is basically the same as BC, but with much higher resolution with 18 layers in the vertical direction. The experiment is carried out in two stages. The coarse grid model ($1°$ north-south $× 1.2°$ east-west) is integrated until a near steady state solution is achieved, then the fine grid model ($1/3° × 0.4°$) is initialized and integrated further.

In Fig. 6-2 the coarse (left) and fine (right) grid solutions are compared. The upper two figures illustrate the equivalent surface elevation, and the lower figures potential vorticity evaluated on the density surface of $\sigma = 26.0$ that exhibits properties of ventilated regions. It is seen from the surface elevation, which also shows the stream function pattern, that the greatest difference in the geostrophic flow field between the two solutions is in the outflow region of the western boundary current. The current in the fine grid solution is much more concentrated in a narrow latitude band. The pattern of the potential vorticity q in the coarse grid case shows agreement with the ventilation theory: the subtropical gyre is dominated by two sources of high q, one at the western boundary and one at the eastern boundary, divided by a tongue of low q water being advected into the interior from the outcrop. In the fine grid case its gradients across the ventilated flow region in the south central and western part of the gyre are eliminated, indicating enhancement of mixing potential vorticity. The analysis shows that the mixing is due to eddies caused by baroclinic instability. The western region of high potential vorticity was not completely included in Luyten *et al.* (1983), which did not take into account the western boundary current. Numerical model studies such as above are effective for comprehensive interpretation, since the analytical solution is difficult to obtain.

In old models using coarse grids, the effect of mesoscale eddies, which are abundant in the ocean, are implicitly approximated by eddy viscosity or eddy diffusion. In high resolution (eddy-resolving) models with grid intervals of less than $1/2°$, their effect can be explicitly described so that we do not have to rack our brains for parameterization procedure, but a huge amount of computer memory and computer time will be needed. For this reason, at first only the limited oceans such as the Gulf Stream extension region (Semtner and Mintz, 1977) or the Atlantic were considered, but in the 1990s it has been made

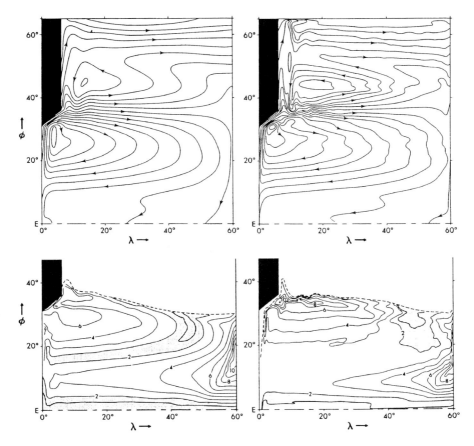

Fig. 6-2. Equivalent surface elevation (top) and potential vorticity on the $\sigma = 26.0$ isopycnal surface (bottom) for the coarse grid case (left), the fine grid case (right) (Cox, 1985).

possible to perform a global eddy-resolving model study (Semtner and Chervin, 1993).

6.2.3 Spin-up of oceanic circulation

a) Spin-up of Rossby waves

In ocean dynamics, a process in which an external force is acted on a system and the responding system forms a new steady state, is called spin-up, and the required time is called spin-up time or response time. Here we consider an example of the ocean in mid latitudes. Figure 6-3 shows a pressure field (equivalent to displacement of the interface in a two-layer model) as a function of longitude (eastward to the right) at different times (marked 1–15) after the

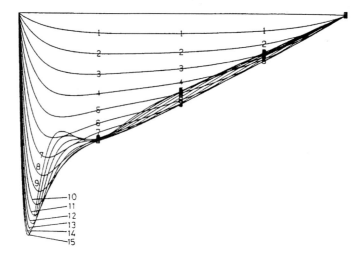

Fig. 6-3. Interface displacement at different times (marked 1 to 15) after the wind with constant negative curl turns on at time zero over a two layer ocean initially at rest. Left and right ends correspond to the western and eastern boundary of the ocean, respectively (Anderson and Gill, 1975).

wind stress is activated at time zero. The figure is obtained by numerically solving the one-dimensional vorticity equation (Anderson and Gill, 1975)

$$
\underset{\begin{array}{cc}(d) & (c)\end{array}}{\frac{\partial\left(\dfrac{\partial^2 h}{\partial x^2} - r^2 h\right)}{\partial t}} + \overset{(b)}{\beta \dfrac{\partial h}{\partial x}} = \overset{(a)}{\dfrac{1}{f_0 \lambda^2} A} \tag{6-14}
$$

where h is the thickness of the lower layer, $r^2 = l^2 + \lambda^{-2}$, l the north-south wavenumber, λ the internal radius of deformation, f_0 the Coriolis parameter, β the planetary beta, and A is the wind stress vorticity assumed to be constant.

Since in the internal region there is no initial pressure gradient ($(b) = 0$), h increases linearly ((a) is nearly equal to (c)) by external forcing (downward flow due to negative wind vorticity). Long internal Rossby waves, moving out from the eastern boundary, establishes a Sverdrup balance with the pressure field as soon as they arrive, and then h stops increasing ($(a) = (b)$). In the west, a western boundary current is formed where the term (d) is significant, but the basic spin-up is completed when the waves reach the western end.

In the center of the subtropical gyre near 30°N, the phase velocity of the first baroclinic mode of Rossby wave is approximately 5 km/day. This means it takes about 2000 days to cross the whole distance 10,000 km of the Pacific

Ocean. Accordingly the response time is several years. Thus the wind-driven circulation discussed in Section 6.2.1 will require this order of the spin-up time to accomplish a steady state. Thermal forcing will also need the same response time scale if the disturbance is added to an existing stratification. However, when the formation of deep water or global circulation including deep-water flow is considered, the time scale in which thermal dissipation will affect the whole system must be taken into account.

b) Response to periodic forcing

Figure 6-4 shows how the equatorial ocean responds to a periodic forcing (Philander and Pacanowski, 1981). East-west components of the current velocity and temperature are plotted as a function of longitude and time at a depth of 112.5 m for two cases of wind forcing periods, 25 days (upper panel, representative of periods between 10 and 50 days) and 200 days (lower panel, representative of periods longer than 50 days). This type of diagram with time and space coordinates is most often used for analyzing time variation.

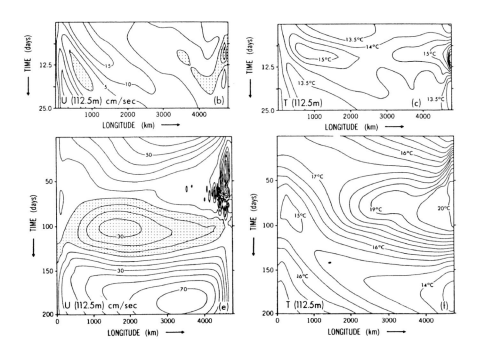

Fig. 6-4. Longitude–time sections of zonal velocity (left) and temperature (right) at the depth of 112.5 m along the equator for the 7th cycle in response to the periodic zonal wind stress with 0.5 dyne/cm² in amplitude. The wind stress period is 25 days (top) and 200 days (bottom) (Philander and Pacanowski, 1981).

The response time for the equatorial zone is, unlike for mid latitude regions, the time it takes for a baroclinic Kelvin wave to propagate eastward across the basin plus the time for the first mode of the reflected Rossby wave to travel back westward across the basin (Philander, 1990). This will be approximately 120 days for the example given in Fig. 6-4 (in which the model width is 5000 km). Thus, the propagation of the phase is significant for the 25-day period forcing, whereas the phased difference is rather small between east and west for the case of a 120-day period. Note that an increase in period is accompanied by an increase in the amplitude of fluctuations and in the depth to which large fluctuates extend. The amplitude is smaller for the shorter period because there is not enough time for the ocean to respond. Another change with increasing period is the phase difference between the wind and surface jets. At shorter periods, the nonlinear eastward jet persists for a considerable time after the onset of westward winds, but at longer periods the persistence of the eastward decreases. It is also to be noted from the figure that winds with a zero mean value give rise to a mean eastward current due to a nonlinear effect.

From the above discussion, it is known that the response to seasonal forcing (one-year period) in mid-latitudes with several years response time, is quite different from that in the equatorial ocean with intra-seasonal response time.

6.2.4 Simulation of El Niño

Since the 1980s, the development of computer capacity, progress of numerical models and increase of available data for exercising calculation and verification have made possible full-scale numerical simulations. The results of calculations have been verified against observed data. Numerical studies designed mainly for dynamics are conducted in the framework of the theory and the prototype dynamical model, while the simulation here implies "numerical observation" which can be obtained only in models. Real-time ocean watching by data assimilation is along this line. There are several reasons why the first full-scale simulation was made in the tropics. More high quality data are available than in mid and high latitudes. ENSO (El Niño/Southern Oscillation) attracted considerable attention. In addition, the response time in low latitudes is shorter.

Philander and Siegel (1985) made an experiment by simulating El Niño of 1982–1983. Their model extends from 130°E to 70°W and 28°S to 50°N latitude. The longitudinal resolution is a constant 100 km, but the latitudinal grid size is 33 km between 10°S and 10°N and increases gradually poleward. The spacing at 25°N is 200 km. The flat-bottom ocean is 4000 m deep. There are 27 levels in the vertical; the upper 100 m have a resolution of 10 m.

On the other hand in the tropical Pacific, TOGA-TAO buoy network was maintained and the measurements were made during the 1982–1983 El Niño (Halpern, 1987). The observed results are illustrated in Fig. 6-5.

The results of the numerical model are shown in Fig. 6-6 to be compared with Fig. 6-5. Temperature variation is in good agreement with the observation: the thermocline drops and sea surface temperature rises around August in 1982,

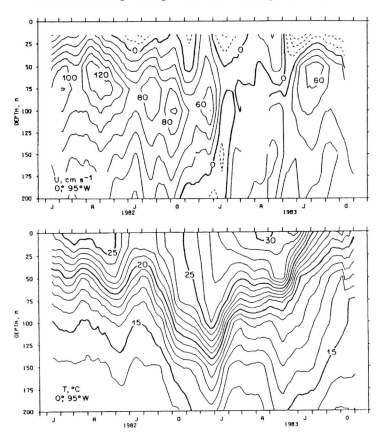

Fig. 6-5. Changes in the zonal velocity (cm/s, top) and in the temperature (°C, bottom) over the upper 200 m on the equator at 95°W observed during 1982–1983 (Halpern, 1987).

with the former abruptly going up at the beginning of 1983 while the latter reaches a maximum in April followed by a sudden fall. Comparison of the vertical structure of the flow also demonstrates the ability of the model to reproduce observed changes: the Equatorial Undercurrent gradually deepens after it attains the greatest velocity around April 1982. It has a maximum velocity in September and disappears for a few months from January 1983, reestablishing its depth as El Niño ends. The rise and fall of the Equatorial Undercurrent correspond to those of the thermocline, reflecting the east-west change of pressure gradient. Therefore, the growth of the Equatorial Undercurrent in the entire tropical Pacific is associated with the onset of El Niño, although the timing might vary location by location; this can be verified by the model result.

The above discussions lead to a conclusion that the simulated results fairly well reproduce variations in the real ocean. Thus, by analyzing the data from the

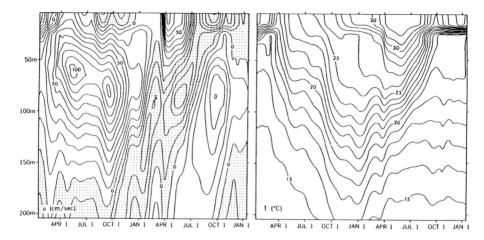

Fig. 6-6. Same as Fig. 6-5 but for the model simulation result (Philander and Siegel, 1985).

simulation, we are able to clarify oceanic variations associated with El Niño, such as the generation and dissipation of equatorial waves, the balance of sea surface temperature change, north-south heat transport in the upper layer, etc.

Although the usefulness of numerical simulation was stressed, its accuracy is not as good as the accuracy of observed data. Improvement of the model performance and the higher input data quality are essential for more precise simulation.

6.3 MODELING OF ABYSSAL CIRCULATION IN DEEP LAYERS

A deep layer in this section means an oceanic layer below 2000 m where global thermohaline circulation is predominant. Very little effect of surface wind reaches deep layers so that currents are rather weak (1 cm/sec in internal regions, not exceeding 10 cm/sec in boundary regions), with the exception of tidal currents and mesoscale eddies. Physical processes are relatively simple, governed by the advection and diffusion of temperature and salinity, except in the limited areas where deep and bottom waters are generated or in the vicinity of mid-oceanic ridges where the earth crust generates thermal flow.

However, abyssal circulation in deep layers is naturally restrained by oceanic topography quite extensively. In particular, bottom water spreads along lower bottom topography. Even at a depth where the topography does not directly affect flow as an obstruction, the water tends to flow parallel to the bottom contours. Therefore, in order to accurately simulate circulation in deep layers, the bottom topography should be exactly expressed first, and then the equations of motion and conservation of temperature and salinity should be solved.

6.3.1 Formation of global abyssal circulation

The theory of abyssal circulation was first given by Stommel (1958), followed by Stommel and Arons (1960). Their theory does not include the effect of bottom topography. Figure 6-7 illustrates the result. It is seen that deep waters over the entire ocean have only two sources: off Greenland in the North Atlantic and the Weddell Sea (heavy black circles in the figure). The water generated in the former area flows south as North Atlantic Deep Water to merge into the Antarctic Circumpolar Current (ACC). The latter (Weddell Sea Deep Water) is also taken into ACC, although part of it goes north in the Atlantic. These waters, after mixing, form a characteristic water mass in the lower part of ACC. Part of this water mass moves northward along western boundaries in the Indian and the Pacific Oceans, filling their deep and bottom layers. On the other hand, everywhere in the ocean except the two source areas, weak upwelling (or vertical stretching of water columns) should occur at the upper boundary of deep layers to compensate the incoming deep and bottom waters. Generally in the internal ocean aside from the western boundary regions, the following vorticity balance (often called the Sverdrup relationship) on the large scale holds

$$\beta v = f \frac{\partial w}{\partial z} \qquad (6\text{-}15)$$

where β (>0) is the derivative of the Coriolis parameter f with distance northward, v the south-north current velocity, w the vertical velocity, and z is the vertical

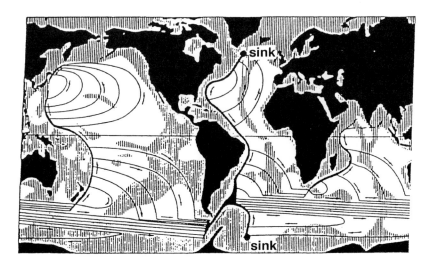

Fig. 6-7. Conceptual figure of global abyssal ocean circulation (Stommel, 1958).

coordinate. Then the vertical stretching of a water column $\partial w/\partial z > 0$ implies poleward internal flow v (northward in the Northern Hemisphere), so that we can expect cyclonic (counterclockwise in the Northern Hemisphere and clockwise in the Southern Hemisphere) circulations as seen in Fig. 6-7.

Figure 6-8 shows the distribution of dissolved oxygen near the ocean bottom (Mantyla and Reid, 1983). Since there is a northward flow in the Atlantic bottom layer, above which the North Atlantic Deep Water flows southward, this figure does not necessarily correspond to Fig. 6-7, but in the Pacific and the Indian Oceans, bottom water with relatively high concentration of oxygen flowing northward is apparent. However, in the Indian Ocean two northward routes exist in the east and west, and unlike in Fig. 6-7 no evident cyclonic circulations can be seen.

6.3.2 Modeling of abyssal circulation

a) Effect of bottom topography on currents

When we consider the effect of bottom topography on currents, the first thing that comes to mind will be its local influence. That is, "currents flow along the topography", "currents pass through a straight", "currents are blocked by an ocean ridge", etc., all include some local influence. This corresponds physically to "large-scale oceanic currents locally flow along geostrophic contours (isopleths of f/H, f: Coriolis parameter, H: characteristic water depth)". Such control by topography is not only limited to the same depth as the topography, but it also extends over shallow currents which the topography cannot affect as an obstacle depending on stratification of the water (stratified Taylor–Proudman's column).

When a current passes through a strait, it is affected not only by the local topography. There should be a downstream flow at the depth of the basin, and the

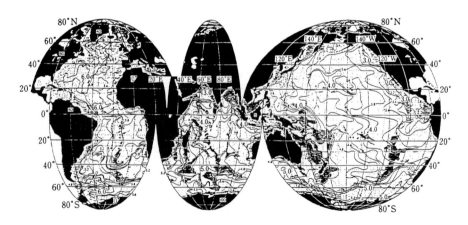

Fig. 6-8. Distribution of dissolved oxygen near bottom (Mantyla and Reid, 1983). Unit is ml/l.

circulation in the basin is governed by the topography of the strait (as in Drake Strait). In general, global thermohaline circulation is driven by the high-pressure field produced by high-density bottom water resulting from sea ice growth and sea surface cooling. Thus, the topography of a strait or an oceanic ridge, by means of passing and stopping the bottom layer flow, restrains how deeply and how far the pressure field associated with thermohaline circulation can propagate from upstream to downstream.

b) Hypsometry effect and the North Pacific deep layer model

We discuss another kind of topographic effect on large-scale circulation. The horizontal area of the cross section of an ocean basin normally increases as the depth decreases. Let us consider such a basin with upwelling in the entire layer (Fig. 6-9), and assume that the vertical velocity is horizontally homogeneous at each level. Then due to the difference of the area, the upward flow is stronger at the lower boundary than at the upper boundary. That is, as a water column goes up, it vertically shrinks ($\partial w/\partial z < 0$) while horizontally expanding. Therefore, the interior flow has an equatorward (southward in the Northern Hemisphere) component according to Eq. (6-15). This, in turn, means a poleward (northward in the Northern Hemisphere) western boundary current to compensate the interior flow, resulting in an anticyclonic gyre. This hypsometric effect was first discussed by Rhines and MacCready (1989).

Ishizaki (1994a, b), by a numerical model study, pointed out the important role of the hypsometric effect in the abyssal circulation in the North Pacific Ocean. That is, the stretching effect of a water column due to the bottom water influx is nearly in balance with the shrinking effect by hypsometry ($\partial w/\partial z = 0$), and an eastward interior flow and a northward western boundary flow are generated (Fig. 6-10). Also in the lower deep layer the hypsometric effect dominates to make an anticyclonic gyre over the North Pacific.

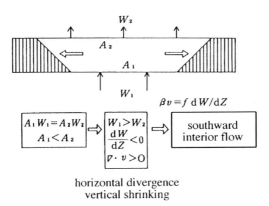

Fig. 6-9. Conceptual figure of the hypsometric effect (Ishizaki, 1994b).

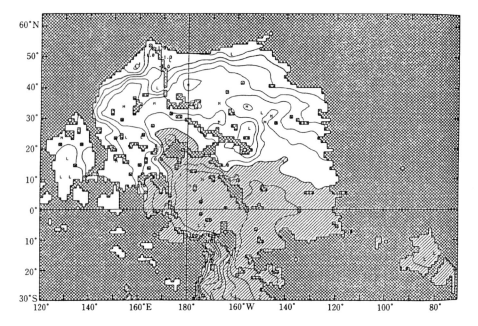

Fig. 6-10. Model pressure field in the bottom layer (4500 m) of the North Pacific (Ishizaki, 1994a). Contour interval is 0.2 cm. Regions with relatively higher pressure are hatched.

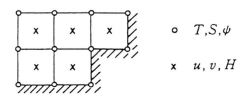

Fig. 6-11. Horizontal grid spacing.

6.3.3 Finite difference scheme for abyssal circulation modeling

The equation of continuity is most important in modeling real abyssal circulations. How it is approximated as a finite-difference equation, will be described below.

Figure 6-11 shows the commonly used horizontal grid point alignment (B-gridpoint). Coastlines are defined by connecting TS grid points (○ in the figure) for temperature (T), salinity (S) and transport stream function (ψ). Velocity components u and v are calculated at UV grid points (× in the figure) positioned at the center of four surrounding TS grid points.

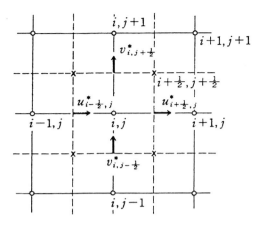

Fig. 6-12. Definition of mass fluxes u^* and v^*.

We now express the continuity equation in two ways: one for TS boxes shown by broken lines in Fig. 6-12 and the other for UV boxes shown by solid lines. The equation for the TS box (i, j) is defined as

$$MC_{i,j}^{T} \equiv u_{i-1/2,j}^* - u_{i+1/2,j}^* + v_{i,j-1/2}^* - v_{i,j+1/2}^* + WF_{i,j,k+1}^{T} - WF_{i,j,k}^{T} = 0 \quad (6\text{-}16)$$

where

$$u_{i-1/2,j}^* = \frac{1}{2}\left(u_{i-1/2,j-1/2}\Delta z_{i-1/2,j-1/2} + u_{i-1/2,j+1/2}\Delta z_{i-1/2,j+1/2}\right)\Delta y$$

$$(6\text{-}17)$$

$$v_{i,j-1/2}^* = \frac{1}{2}\left(v_{i-1/2,j-1/2}\Delta z_{i-1/2,j-1/2} + v_{i+1/2,j-1/2}\Delta z_{i+1/2,j-1/2}\right)\Delta x$$

and WF^{T} is the vertical mass flux; Δx, Δy, and Δz are two horizontal and vertical grid intervals, respectively. If this layer touches the bottom, Δz is horizontally variable. The vertical suffix $(k + 1/2)$ is omitted in the equations. The boundary condition at the sea surface and bottom will be: $WF^{T}_{i,j} = 0$. The equations can be applicable even when any of the surrounding UV boxes happens to be on land, by making Δz zero. As is illustrated in Fig. 6-13(a), this expression of the continuity equation is naturally defined and can be applied to baroclinic flow without forfeiting physical constraints due to topography, and also to vertical distribution of vertical velocity components (Suginohara, 1978; Ohnishi, 1978).

The continuity equation for a UV box $(i + 1/2, j + 1/2)$ is derived from equations for surrounding TS boxes

$$MC^{U}_{i+1/2,j+1/2} \equiv \frac{MC^{T}_{i,j}}{N_{i,j}} + \frac{MC^{T}_{i+1,j}}{N_{i+1,j}} + \frac{MC^{T}_{i,j+1}}{N_{i,j+1}} + \frac{MC^{T}_{i+1,j+1}}{N_{i+1,j+1}} \qquad (6\text{-}17')$$

where $N_{i,j}$ is the number of sea boxes around the TS point (i, j). As long as the UV box under consideration is sea, each of Ns in the equation is nonzero. Then, the vertical mass flux for the UV box $(i + 1/2, j + 1/2)$, $WF^{U}_{i+1/2, j+1/2}$, is defined by surrounding WF^{T}'s as

$$WF^{U}_{i+1/2,j+1/2} = \frac{WF^{T}_{i,j}}{N_{i,j}} + \frac{WF^{T}_{i+1,j}}{N_{i+1,j}} + \frac{WF^{T}_{i,j+1}}{N_{i,j+1}} + \frac{WF^{T}_{i+1,j+1}}{N_{i+1,j+1}}. \qquad (6\text{-}18)$$

In this definition, $WF^{U}_{i+1/2,j+1/2}$ at the bottom can have a nonzero value if any one of the adjacent UV boxes has a greater layer number than that at $(i + 1/2, j + 1/2)$, naturally expressing oblique upward or downward flows along the bottom slope.

Figure 6-13 illustrates the distributions of WF^{T} (a) and WF^{T} (b) for barotropic current flowing over a step-like bottom topography (Ohnishi, 1978). According to Eq. (6-18), a discontinuity will occur between the top surfaces of UV boxes A and B, and also the bottom of A and B will have nonzero WF^{U}, which would not be physically acceptable. These discrepancies will be dissolved by introducing oblique upward fluxes along the bottom slope as shown in the figure. Physical properties of vertical velocity will also be conserved. Mass fluxes and momentum advection using those fluxes are discussed in detail by Takano (1978a, b) and Ohnishi (1978).

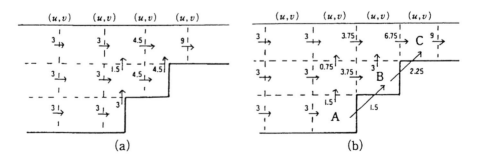

Fig. 6-13. Distribution of vertical mass fluxes WF^{T} (a) and WF^{U} (b) for a barotropic flow passing over stair-like topography (Ohnishi, 1978).

6.4 MODELING OF SEA ICE

Sea ice heavily affects the ocean–atmosphere interaction in high latitudes. For instance, since its thermal insulation capacity is high, sea ice restrains heat exchange between the ocean and the atmosphere. Sea ice, while moving (drifting), acts upon momentum exchange between the two fluids. For this reason, modeling of sea ice is important for numerical model studies on the ocean–atmosphere interaction in high latitudes.

6.4.1 History of sea ice modeling

Modeling of sea ice was developed as thermodynamic models and dynamical models. Maykut and Untersteiner (1971), in their pioneering work, proposed a one-dimensional thermodynamic model in which snow fall, salinity distribution in sea ice, internal melting due to solar radiation, etc. were taken into account for perennial ice in the Arctic. Semtner (1976) simplified this model for his climate model study using a three-level snow-ice system. Parkinson and Washington (1979) combined Semtner's model with a simple dynamical model to study seasonal variations of sea ice in the Arctic and Antarctic regions. Hibler (1979) was the first to thoroughly discuss the dynamical processes of sea ice by introducing the concept of rheology into the internal stress, and his model succeeded in describing the distribution of sea ice thickness in the Arctic and its motion.

In recent years, coupled ice–ocean models were developed by Hibler and Bryan (1987), Semtner (1987), Mellor and Kantha (1989), Piacsek et al. (1991) and Oberhuber (1993). In their ocean model Mellor and Kantha, Piacsek et al. embedded the turbulence closure model of Mellor and Yamada (1974, 1982) to express the mixing layer, which plays an important role in the coupled ice–ocean system. Häkkinen and Mellor (1992) simulated the Arctic ice–ocean system using the Mellor–Kantha model. The Antarctic region has been studied by a coupled ice-mixing layer model (Stössel and Lemke, 1990).

In this section the coupled ocean–atmosphere model by Mellor and Kantha (1989) will be described in detail. Then the results of Häkkinen and Mellor (1992) will be introduced to understand the targets of sea ice-modeling research.

6.4.2 Coupled ice–ocean model

The ocean in high latitudes is considered to consist of sea ice covered with snow and the lead (waterway) as shown in Fig. 6-14. In the ice model the equations of ice motion, the mass conservation equation, and an empirical formula for ice concentration (the ratio of the sea ice area to the open water) are solved to obtain the sea ice velocity (U_{Ii}, $I = x$, y), thickness (h_I) and ice concentration (A). At the same time, the momentum flux, heat flux and salinity flux can be obtained. The ocean is driven by these fluxes and the coupled ice–ocean system is completed.

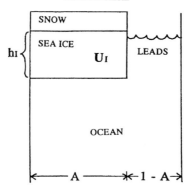

Fig. 6-14. Components in the model of the polar ocean. Sea-ice drift velocity (U_1), thickness (h_1) and concentration (A) are indicated.

a) Ice model

The dynamic equation applicable to open water ($A = 0$), an ensemble of ice floes ($A < 1$), or for solid ice ($A = 1$) is

$$\frac{\partial(Ah_1U_{1i})}{\partial t} + \frac{\partial(Ah_1U_{1j}U_{1i})}{\partial x_j} = Ah_1\varepsilon_{ijk}f_jU_{1k} - Ah_1g\frac{\partial \zeta}{\partial x_i} + \frac{\left(\frac{\partial \sigma_{ij}}{\partial x_j}\right)}{\rho_1} + A\frac{\tau_{A1i} - \tau_{1Oi}}{\rho_1}$$

(6-19)

where $f_j = (0, 0, f)$ and f is the Coriolis parameter; ρ_1 is the density of ice; σ_{ij} is the internal ice stress tensor; τ_{A1i} is the atmospheric wind stress; τ_{1Oi} the ice–ocean interfacial stress; and g is the gravity acceleration constant (Fig. 6-15). The first term on the right-hand side of Eq. (6-19) corresponds to the Coriolis force, the second term the pressure gradient, the third term the internal stress, and the forth term the stress force at the ice–ocean interface.

The atmospheric wind stress τ_{A1i} is given by

$$\left(\tau_{A1x}, \tau_{A1y}\right) = C_{DA1}\rho_a|U_{10}|\left(U_{x10}, U_{y10}\right)$$

(6-19')

where C_{DA1} is the bulk coefficient; ρ_a is the atmospheric density; $U_{10} = (U_{x10}, U_{y10})$ is the wind vector at a height of 10 m.

The ice–ocean interfacial stress τ_{1Oi} is given by

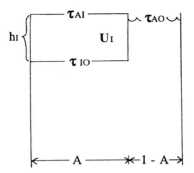

Fig. 6-15. Wind stress (τ_{AI}), water stress (τ_{IO}) on sea ice and wind stress (τ_{AO}) on leads.

$$\frac{\left(\tau_{IOx}, \tau_{IOy}\right)}{\rho_o} = \frac{ku_\tau\left(U_{Ix} - U_x, U_{Iy} - U_y\right)}{\ln\left(\dfrac{z}{z_0}\right)}, \quad z \to 0 \qquad (6\text{-}19'')$$

where ρ_O is the density of the ocean; k is the Karman constant; u_τ is the friction velocity $u_\tau = (\tau^2_{IOx} + \tau^2_{IOy})^{1/4}\rho_0^{-1/2}$; $U = (U_x, U_y)$ is the current vector in the first layer; z_0 is the roughness parameter.

The internal stress of ice is expressed based upon Hibler (1979) as

$$\sigma_{ij} = 2\eta_0\varepsilon_{ij} + \left[(\zeta_0 - \eta_0)\varepsilon_{kk} - \frac{P_1}{2}\right]\delta_{ij} \qquad (6\text{-}19''')$$

where ε_{ij} is the strain velocity defined by

$$\varepsilon_{ij} = \frac{\dfrac{\partial U_{Ii}}{\partial x_j} + \dfrac{\partial U_{Ij}}{\partial x_i}}{2} \qquad (6\text{-}19'''')$$

and η_0 is the shear viscosity; ζ_0 is the bulk viscosity.

The equation for the conservation of the mass of ice is

$$\frac{\partial(Ah_1)}{\partial t} + \frac{\partial(Ah_1 U_{Ii})}{\partial x_i} = \frac{\rho_0\left[A(W_{IO} - W_{AI}) + (1-A)W_{AO} + W_{FR}\right]}{\rho_1}. \qquad (6\text{-}20)$$

The empirical equation for the ice concentration is

$$h_1 \left[\frac{\partial A}{\partial t} + \frac{\partial (AU_{Ii})}{\partial x_i} \right] = \frac{\rho_o \left[\Phi(1-A)W_{AO} + (1-A)W_{FR} \right]}{\rho_1}, \quad 0 \le A \le 1 \quad (6\text{-}21)$$

where W_{AI} is the melting rate (positive) on the top of the ice: it is also the freezing rate (negative) when trapped surface water refreezes in late summer. W_{IO} is the freezing rate (positive) of congelate ice at the ice–ocean interface. W_{AO} is the melting (negative) rate or freezing rate (positive) in open water. W_{FR} is the freezing rate of frazil ice in the sea water. Figure 6-16 illustrates the schematic arrangement of these volumetric fluxes. Φ is an empirical constant.

b) Coupling of ocean model and ice model
      The boundary conditions for momentum, heat and salinity at the sea surface are

$$K_M \left(\frac{\partial U_x}{\partial z}, \frac{\partial U_y}{\partial z} \right) = A \frac{\left(\tau_{IOx}, \tau_{IOy} \right)}{\rho_o} + (1-A) \frac{\left(\tau_{AOx}, \tau_{AOy} \right)}{\rho_o}, \quad z \to 0 \quad (6\text{-}22)$$

$$\rho_o C_{pO} K_H \frac{\partial T}{\partial z} = -F_T, \qquad\qquad\qquad\qquad z \to 0 \quad (6\text{-}23)$$

$$K_H \frac{\partial S}{\partial z} = -F_S, \qquad\qquad\qquad\qquad\qquad z \to 0 \quad (6\text{-}24)$$

where U is the velocity, T is the temperature, and S is the salinity. τ_{AO} is the wind stress acting on the sea surface in the lead (open ocean) as shown in Fig. 6-15.

$$\left(\tau_{AOx}, \tau_{AOy} \right) = C_{DAO} \rho_a |U_{10}| \left(U_{x10}, U_{y10} \right). \qquad (6\text{-}24')$$

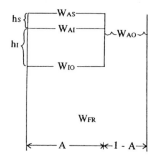

Fig. 6-16. Locations of variables for volume fluxes.

With the above three conditions, the equations for momentum, the conservation of heat, and the conservation of salinity are to be solved for the coupled ice-ocean system, of the coupled ocean. General methods of solving coupled models will be discussed in Section 6.5.

6.4.3 Simulation of the Arctic region by a coupled ice–ocean model

Häkkinen and Mellor (1992) conducted a simulation study on seasonal variations of the ice–ocean system in the Arctic Ocean and its peripheral seas, using the coupled ice–ocean model described in the previous section. Their ocean model was initialized with the annual average hydrographic climatology of Levitus (1982) with sea ice concentration by Parkinson *et al.* (1987). River runoff and atmospheric data over the ocean were used as driving forces. Integration for 15 years led to an equilibrium solution with seasonal cycle. The results are shown below.

Figure 6-17 depicts seasonal evolution of the ice thickness and concentration fields in May (a) and August (b). It can be seen that the seasonal ice extent variations are realistic, with the exception of a small area in the central Greenland Sea, where the ice cover tends to extend too far east. The simulated minimum ice cover retreats to about the average observed ice cover (Parkinson *et al.*, 1987). The ice thickness also conforms to the limited set of observed values (Bourke and Garrett, 1987).

Figure 6-18(a) shows the annual average ice motion, which is a very smooth field due to the smooth climatological forcing but shows the main features of the

(a) MAY   (b) AUGUST

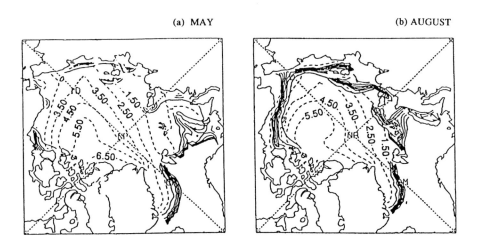

Fig. 6-17. Sea ice concentration (solid lines, larger than 30% with 5% interval) and thickness (broken lines, 1 m interval) of May and August in the model (Häkkinen and Mellor, 1992). Concentrations of sea ice thinner than 10 cm are not contoured.

a   **b**

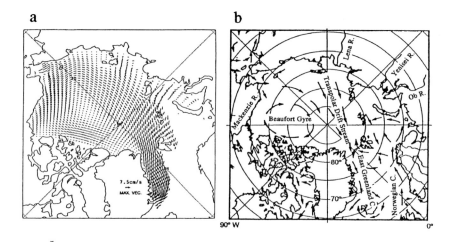

Fig. 6-18. (a) Annual mean sea ice drift vectors in the model (Häkkinen and Mellor, 1992). (b) Schematic view of sea ice drift and current from observation (b). Vector magnitudes larger than 7.5 cm s^{-1} in the model are truncated.

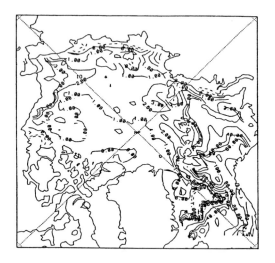

Fig. 6-19. Annual mean oceanic heat flux F_T in the model. Contour intervals are 1 W m^{-2} for –4 to 4 W m^{-2} and 20 W m^{-2} for values above 20 W m^{-2} (Häkkinen and Mellor, 1992).

Beaufort Gyre and the Trans-Polar Drift Stream. In Fig. 6-18(b), a schematic view of observed surface currents and ice drift, is shown .

Figure 6-19 shows the annual average oceanic flux (F_T). The net heat loss from the deep ocean in the central basin is well below 2 W m^{-2}, and in most of the area below 1 W m^{-2}. Overall the central Arctic heat loss is small because of the existence of the strong upper ocean stratification in the main Arctic. On the other hand, large heat losses occur in the Greenland Sea open water area, 40–120 W m^{-2}.

Among the ocean model results, Fig. 6-20(a) illustrates the summer (mid-September) surface salinity. The simulated pattern and values are in reasonably good agreement with the observed surface salinity field shown in Fig. 6-20(b). At least the central basin salinity is within 0.5–1 psu of the observed values with a more severe degradation toward the coasts, particularly near the mouths of the McKenzie, the Lena, the Enesey, and the Obi rivers. One probable cause for the high coastal salinity values is that the river runoff is mixed to the total water column instead of to a thin surface layer.

The vertical cross sections for temperature (T), salinity (S) and potential density (σ) are shown in Fig. 6-21. The upper panels (a) show the observed climatology data used for initial conditions, and the lower panels (b) the simulated results at the end of year 15. While the T and S fields have changed considerably in the Eurasia Basin and the Greenland Sea, the changes compensate each other to produce a similar density field. The Canada Basin shows a cooling of the Atlantic origin water, and a weak relaxation of the strong halocline in the upper 200 m, while the rest of the density structure remains nearly the same.

The oceanic velocity fields in the end of December of year 15 at depths of 50 m and 1000 m are shown in Fig. 6-22. The near-surface circulation (a) shows the dominant oceanic counterparts of the Beaufort Gyre, the Transpolar Drift Stream, the East Greenland Current, and the Norwegian Current. Compared to observations (see Fig. 6-18(b)), the simulated Drift Stream, which is basically

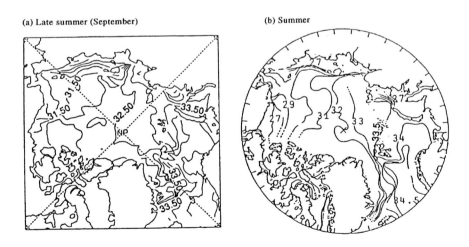

(a) Late summer (September)   (b) Summer

Fig. 6-20. Sea surface salinity for (a) late summer in the model (0.5 psu interval) and for (b) summer from observation (Häkkinen and Mellor, 1992).

(a) Observation

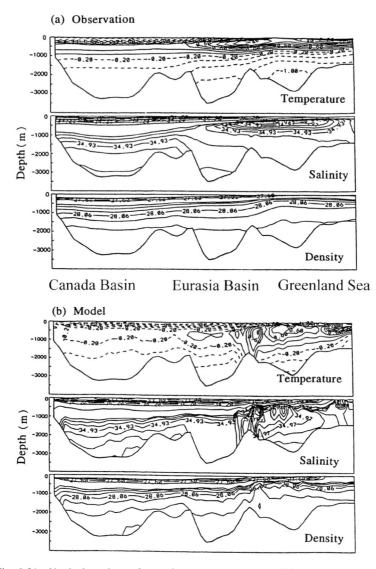

Fig. 6-21. Vertical sections of annual mean temperature, salinity and density in the
Greenland Sea, Eurasia Basin and Canada Basin (Häkkinen and Mellor, 1992). Contour
intervals are 0.2°C for temperature, 0.2 psu for salinity above 34.8 psu and 0.01 for
salinity below 34.9 psu, and 0.2 for density above 28.0 (σ_t) and 0.01 for density below
28.0.

wind driven, appears to be too far to the east, but other patterns are fairly well
reproduced. At 1000 m (b), the anti-cyclonic gyres in the Canada Basin and in the
western half of the Eurasia Basin are more prominent than the flow along the
boundaries.

(a)   50m depth   (b)   1000m depth

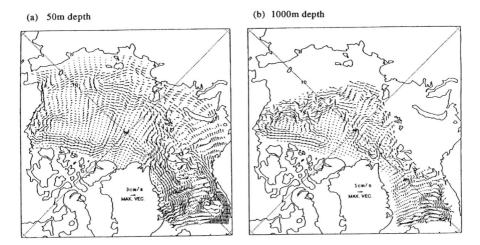

Fig. 6-22. Currents at (a) 50 m depth and 1000 m depth in the model (Häkkinen and Mellor, 1992). Vectors are truncated for values larger than (a) 3 cm s^{-1} and (b) 1 cm s^{-1}.

6.5 MODELING OF CLIMATE CHANGE
BY COUPLED OCEAN–ATMOSPHERE MODELS

Researches on ocean–atmosphere interactions have long been carried out under two categories: (1) How does the ocean respond to atmospheric and climatological variations, such as winds over the sea? (2) How does the atmosphere respond to varying sea surface temperatures? Numerical modeling for the ocean used to aim at obtaining an answer for the question (1). Thanks to today's high-speed powerful computers, an ocean–atmosphere coupled general circulation model can be executed where an oceanic and an atmospheric model are run simultaneously while calculating the interactions (vertical fluxes). In this chapter the methodology of the coupled model will be mainly discussed. For detailed discussion on this topic, see Tokioka *et al.* (1993).

6.5.1 Methods for running coupled ocean–atmosphere models

a) Contact point of atmospheric models and ocean models
Atmosphere contacts land on the earth and seawater on the sea surface. Various physical quantities are exchanged through the sea surface (including floating ice), just as through the land surface. This section will deal with physical interactions that are associated with dynamics, but chemical substances such as carbon dioxide, which are not directly related to dynamics, will play a significant role in long-term climate change. In coupled ocean–atmosphere models, vertical fluxes passing through the wall which connects two boxes, one in the ocean, the other in the atmosphere, are directly calculated.

The most important vertical flux in the ocean is that of the horizontal momentum, or the horizontal component of wind stress. It is obtained from the momentum equilibrium relationship of the atmospheric boundary layer, which is normally incorporated in the atmospheric general circulation model within the coupled model. For the atmosphere, it is sea surface friction. If sea ice exists, the stress on ice is determined from its roughness by the atmospheric boundary layer model. At the same time, the stress at the ocean–sea ice boundary is determined by the ice model included in the general ocean circulation model within the coupled model.

Sea surface pressure (sea level) varies as the atmospheric pressure changes, but it is normally neglected because its effect on the general circulation is small compared with that of wind-driven circulation.

The exchange of thermal energy is then estimated in the atmospheric general circulation model. The sensible and latent heat fluxes are calculated by the bulk method using the sea surface temperatures passed on from the oceanic model. Meanwhile, the solar radiation coming into the sea surface and the long wave radiation returning from it are calculated by the radiation equilibrium relationship in the atmosphere. These will determine the amount of heat content which makes the seawater temperature distribution vary. The solar radiation attenuates down to 10–20 meters, as it is gradually converted to internal heat in seawater.

The budget of salt is calculated by precipitation at the sea surface, river run-off, and sea surface temperature in the atmospheric model, combined with the salinity flux in the oceanic model. Cloud vapor and heat sources for cumulus are also estimated using the amount of evaporation in the atmospheric model. Vertical water flux causes change in sea level, or sea surface pressure. This effect, however, is considered small compared with that of the large-scale variation of sea surface pressure induced by wind (the source of the wind-driven circulation), and thus is usually neglected.

In addition to the dynamical quantities, substances are exchanged at the sea surface, too. The present coupled models place more emphasis on dynamical interactions and therefore neglect these, but vertical fluxes of global warming substances such as carbon dioxide should eventually be included in the long-term climate change models. Further, in the coupled models on the paleontological or geological time scale, heat and carbon fluxes to and from the solid earth should be taken into account.

b) Preparation for the coupled ocean–atmosphere model
(i)   Necessary models (component parts) and their capacity
       What kind of models as parts are required for the coupled ocean–atmosphere model, depends on the temporal scale of the interaction under consideration. For climate study in a geological time scale, a carbon cycle model will be needed as a part. For glacial climate, an ice sheet model will be necessary. Here we are going to consider climate variability for 50–100 years, so that we will need an ocean–atmosphere dynamical general circulation model and an ice–land model.

These models as parts must be provided with the following capabilities.

Firstly, every model should reasonably reproduce seasonal variations (or intra-seasonal variations in some cases) of climate. The driving forces for the coupled model are vertical fluxes at the sea surface, but it is not guaranteed that their computed values are correct. Therefore, the climatological boundary conditions, which are believed to be correct, should at least reproduce observed climate values. Secondly, ocean and atmosphere should vary (respond) consistently with observations as historical boundary conditions (fluxes) change.

Without these two capabilities being assured, no correct solution would be expected when the models are coupled. If a physically or observationally unrealistic coupled solution is found even when these conditions are met, then we should reconsider how to obtain the correct vertical fluxes.

Necessary data for a coupled model are in most cases just the sum of the data used for running its parts. However, complete sets of good data are rarely available, especially for sea surface wind stress, sea surface temperature, sea surface salinity, or distributions of vertical heat flux and water flux. Assimilated data in the atmospheric prediction model are essential for detailed comparison of fluxes, since they can be obtained for all grid points, whereas observed data almost always have missing points.

(ii) Method of coupling

Figure 6-23 is a schematic diagram of a coupled model showing its overall structure. Solar radiation from outside of the ocean–atmosphere system would be an external force for an ideal coupled model. In practice, however,

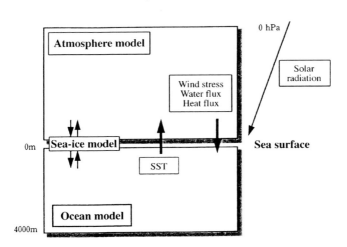

Fig. 6-23. The coupled ocean–atmosphere model is forced by solar radiation. Atmosphere model receives SST data from the ocean model, whereas the ocean model is driven by sea surface wind stress, precipitation–evaporation, and vertical heat flux calculated in the atmosphere model.

flux adjustment (a kind of forcing) will be made using observed climatological values in the ocean and atmosphere. In certain intervals (for instance, every 24 hours) data are to be exchanged between the atmosphere and the ocean. Within that time, the atmospheric and oceanic models run independently of each other. The oceanic model passes surface temperatures to the atmospheric one, which provides the wind stress, precipitation, and vertical heat transport to the former. Solar radiation penetrates into the ocean giving heat with a constant attenuation rate (for instance, $1/e$ for 10 m).

(iii) Initial condition and flux adjustment

In an experiment (e.g., El Niño simulation), which is designed to test the capability of a coupled system with a given natural force such as solar radiation, it takes a few to ten years for the atmospheric model and oceanic model to adapt to each other, that is, to reach a quasi-equilibrium state in coupling (initial climatological values). Therefore, initial conditions at the start of coupled integration are not decisive, unless there are multiple equilibrium solutions in the system. Usually appropriate climatological data observed in the atmosphere and ocean will be satisfactory as initial values. When the variability of more than 100 years in the coupled system is of your interest, calculations for coupled adapting should be performed for several hundred years until the deep ocean reaches a new adjusted state.

If the model is run for a long period of time until the coupling is adapted, the model climate values usually shift from observed values because of incorrect sea surface fluxes. In order to overcome this defect, the following method is normally adopted. First, the ocean model is run to reach a stationary state with fluxes proportional to the difference between a model variable and observation. The fluxes are averaged over several years. Then the coupled system is started and these artificial fluxes are added to those provided by the atmospheric model, the total fluxes being passed on to the ocean model (flux adjustment). In this way, the climatology of seawater temperature and salinity are achieved. However, even after such flux adjustment at the sea surface is made, it often happens that the simulated sea water temperature in the tropics is lower by 1–3°C. One of the most challenging tasks in making a successful coupled model lies in reproducing the seasonal variation of the correct climatology of seawater temperature and its horizontal distribution.

Now forecasting of El Niño by the coupled model is made almost possible half a year or one year beforehand. Nevertheless the model climatology is often poorly predicted and initial values should be carefully selected. See Kitamura (1994) for a detailed overview of the coupled model used for forecasting.

6.5.2 Modeling of climate change by coupled models

a) Modeling of ENSO

The first success for simulating ENSO (El Niño/Southern Oscillation) was achieved in a model where a coupled process was simplified with given basic fields (climatological data) both in ocean and atmosphere. Forecasting ENSO was also successful to a certain degree (see Chapter 5). ENSO simulation and its prediction experiments were also performed using a hybrid model in which a realistic ocean general circulation model was coupled with an atmospheric model with statistics (correlation between SST and sea surface wind). Coupled General Circulation Models (CGCM) were further developed to compare the results with three-dimensional observations in the atmosphere and ocean, and to examine quantitatively the effect of ENSO on the global climate. They succeeded in reproducing interannual variations that resemble the ENSO phenomena with realistic seasonal variations in the atmospheric and oceanic fields. An example of such a coupled model by Nagai *et al.*, (1992) will be introduced in the following.

The oceanic component of the coupled model is a GCM for the Pacific, while the atmospheric one is a global GCM. It was time-integrated for 30 years to study interannual variability in the tropics. Figure 6-24 shows the appearance of a model El Niño event in years 9 and 10. Simulated temperature anomalies averaged vertically in the upper 300 meters are plotted in time sequence. It is seen that the warm anomalies, which are first piled up in the Western Equatorial Pacific, eventually propagate eastward, resulting in the El Niño (October in year 10). At the same time the wind system shifts its phase (see Chapter 5). In this respect, the temperature anomaly pattern in the model well reproduces El Niño, but the absolute values of sea surface temperature do not agree with observations: the simulated temperatures are lower by 2–3°C, weakening the intensity of the ocean–atmosphere interaction (the amplitude of the El Niño).

b) Modeling for global climate change
(i)   Global coupled ocean (ice)–atmosphere model
      The coupled model for studying ENSO does not include the effect of sea ice in high latitudes. The coupled global model with the sea-ice model (see Section 6.4) in it, has been developed mainly for evaluating the influence of chemical substances such as carbon dioxide on global warming. It differs from the ENSO model only in introducing the sea-ice model, but this new component causes the salinity in high latitudes to vary seasonally and interannually (including its long-term drift), affecting global general circulations in deep and bottom layers over a thousand years. Since such inter-annual variations might be possible in real oceans, an attempt to reproduce the climatology in the global coupled model faces a fundamental difficulty. No global coupled model has been successful yet in obtaining climatological values without drift, after completing all necessary coupling adjustments including deep-sea circulation, which would require coupled integration beyond a thousand years.

(ii) Interdecadal variations in the global coupled model
      Coupled numerical models have been integrated over 100 years to investigate global warming effect. As a by-product, long-term climatological and oceanic natural variabilities in the Atlantic and the Pacific have been reported. A 20-year time scale variability, which was detected in the global

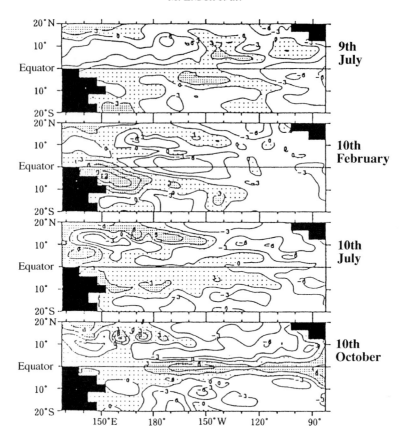

Fig. 6-24. Anomaly of the upper 300 m averaged ocean temperature representing the model El Niño, from the 9th to 10th year of the coupled model run. Hatched warm anomaly is accumulated in the tropical western Pacific and moves eastward. Atmospheric system shows characteristic changes as Southern Oscillation.

coupled model of the Meteorological Research Institute/JMA, will be introduced here. For the corresponding observations, see Chapter 4.

The model is integrated for 70 years. After removing interannual variations on the ENSO time scale by applying a low-pass filter with a cut-off period of 12 years, the time series of physical parameters were decomposed into empirical orthogonal functions (EOF). Figure 6-25 shows the first EOF modes of sea surface temperature (upper panel) and surface wind stress (lower panel), showing their eigenvectors and time coefficients, respectively (Yukimoto *et al.*, 1996). The SST variation shows a wedge-shaped pattern with its apex at the central equator, spatially similar to the ENSO time scale variation. It is noted that in this case the amplitude in the central Pacific is larger than that

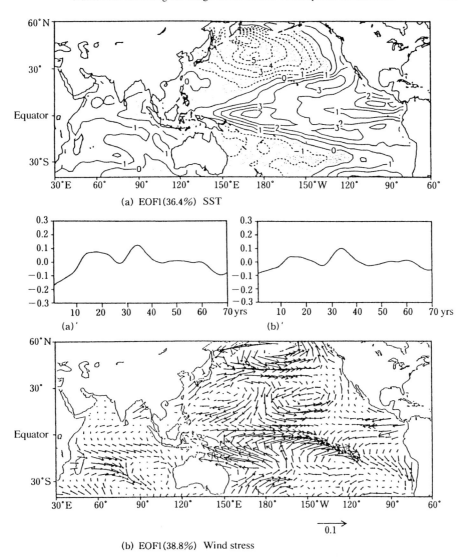

(a) EOF1(36.4%) SST

(a)'

(b)'

(b) EOF1(38.8%) Wind stress

Fig. 6-25. Spatial distribution and time coefficient of the EOF first mode of long term components of Sea Surface Temperature (SST) (a) and sea surface wind stress (b) (Yukimoto *et al.*, 1996). Low pass filtered data (12 years) was analyzed.

in the equatorial region, unlike in the case of the ENSO. Their result indicates that the interdecadal variability found in the mid latitudes of the Pacific is induced by that in the tropics, although some others report that it is caused by coupled interactions in mid latitudes, having little correlation with lower latitudes. Detailed analyses should be made for further clarification on this topic.

(iii) Global coupled model for evaluation of global warming due to carbon dioxide increase

The global coupled model was first developed to evaluate the effect of the carbon dioxide increase upon climate. The time scale of the increase of such artificial warming substances as carbon dioxide is 100 years at most. In the model, carbon dioxide in the atmosphere is initially increased by a realistic value. Then the radiation equilibrium calculated in the atmospheric model changes, resulting in increasing temperature and the global warming starts. However, as mentioned earlier in this chapter, the coupled model contains the long-term climate drift and the natural interdecadal variations of the atmosphere–ocean–ice system. They cannot be separated easily from the artificial effect. Accordingly, a control experiment without increasing the present carbon dioxide under the exactly same initial and boundary conditions should be made to examine the natural variabilities including climate model drift. The difference of the two calculations will provide overall linear effect of carbon dioxide increase. The similar model experiment for evaluating global warming is presently conducted in many countries.

REFERENCES

Anderson, D. L. T. and A. E. Gill (1975): *Deep Sea Res.*, **22**, 583–596.
Anderson, D. L. T. and A. E. Gill (1986):
Bourke, R. H. and R. P. Garrett (1987): *Cold Reg. Sci. Technol.*, **13**, 259–280.
Bryan, K. (1969): *J. Computational Physics*, **4**, 347–376.
Bryan, K. and M. D. Cox (1967): *Tellus*, **19**, 54–80.
Cox, M. D. (1985): *J. Phys. Oceanogr.*, **15**, 1312–1324.
Cox, M. D. and K. Bryan (1984): *J. Phys. Oceanogr.*, **14**, 674–687.
Danabasoglu, G., J. C. McWilliams and P. R. Gent (1994): *Science*, **264**, 1123–1126.
Häkkinen, S. and G. L. Mellor (1992): *J. Geophys. Res.*, **97**, 20285–20304.
Halpern. D. (1987): *J. Geophys. Res.*, **92**, 8197–8212.
Hibler, W. D., III (1979): *J. Phys. Oceanogr.*, **9**, 815–846.
Hibler, W. D., III and K. Bryan (1987): *J. Phys. Oceanogr.*, **17**, 987–1015.
Ishizaki, H. (1994a): *J. Phys. Oceanogr.*, **24**, 1921–1939.
Ishizaki, H. (1994b): *J. Phys. Oceanogr.*, **24**, 1941–1954.
Kitamura, Y. (1994): *Gekkan Kaiyo Kagaku, Gogai No. 6*, Prof. Yutaka Nagata Retirement Commemoration Issue, Physics in Ocean, pp. 74–81 (in Japanese).
Levitus, S. (1982): *Climatologycal Atlas of the World Ocean*, NOAA Publ., No. 13, U.S. Dep. of Comm., Washington, D.C., 173 pp.
Luyten, J. R., J. Pedlosky and H. Stommel (1983): *J. Phys. Oceanogr.*, **13**, 292–309.
Mantyla, A. W. and J. L. Reid (1983): *J. Mar. Res.*, **30**, 805–833.
Maykut, G. A. and N. Untersteiner (1971): *J. Geophys. Res.*, **76**, 1550–1575.
Mellor, G. L. and L. H. Kantha (1989): *J. Geophys. Res.*, **94**, 10937–10954.
Mellor, G. L. and T. Yamada (1974): *J. Atmos. Sci.*, **31**, 1791–1806.
Mellor, G. L. and T. Yamada (1982): *Review Geophys. & Space Phys.*, **20**, 851–875.
Mesinger, F. and A. Arakawa (1976): Numrical Methods Used in Atmospheric Models, ICSU & WMO, GARP Publications Series, #17, 64 pp.
Munk, W. H. (1950): *J. Meteor.*, **7**, 79–93.
Nagai, T., T. Tokioka, M. Endoh and Y. Kitamura (1992): *J. Climate*, **5**, 1202–1233.
Nitta, T., *et al.* (1972): Kisho Kenkyu Note, Numerical Methods for Dynamical Meteorology, Japan Meteorological Society, No. 110 (in Japanese).

Oberhuber, J. M. (1993): *J. Phys. Oceanogr.*, **23**, 808–829.

Ohnishi, Y. (1978): *Oceanography as Environmental Science 2*, ed. by S. Horibe, Tokyo University Press, pp. 246–271 (in Japanese).

Parkinson, C. L. and W. M. Washington (1979): *J. Geophys. Res.*, **84**, 311–337.

Parkinson, C. L., J. C. Comiso, H. J. Zwally, D. J. Cavalieli, P. Gloersen and W. J. Campbell (1987): *Arctic Sea Ice 1973–1976—Satellite Passive Microwave Observations*, NASA Spec. Publ., SP-489, 296 pp.

Philander, S. D. H. (1990): *El Niño, La Niña, and the Southern Oscillation*, Academic Press, 293 pp.

Philander, S. D. H. and R. C. Pacanowski (1981): *J. Geophys. Res.*, **84**, 311–337.

Philander, S. D. H. and A. D. Siegel (1985): *Coupled Ocean—Atmosphere Models*, ed. by J. Nihoul, Elsevier, pp. 517–541.

Piacsek, S., R. Allard and A. Warn-Varnas (1991): *J. Geophys. Res.*, **96**, 4631–4650.

Rhines, P. B. and P. M. MacCready (1989): *Parameterization of Small-Scale Processes*, ed. by P. Mueller, Hawaii Institute of Geophysics Special Publication, University of Hawaii, pp. 75–99.

Robinson, A. R. and P. Welander (1963): *J. Mar. Res.*, **21**, 25–38.

Semtner, A. J., Jr. (1976): *J. Phys. Oceanogr.*, **6**, 379–389.

Semtner, A. J., Jr. (1987): *J. Phys. Oceanogr.*, **17**, 1077–1099.

Semtner, A. J. and R. M. Chervin (1993): *J. Geophys. Res.*, **97**, 5493–5550.

Semtner, A. J. and Y. Mintz (1977): *J. Phys. Oceanogr.*, **7**, 208–230.

Stommel, H. (1948): *Trans. Am. Geophys. Union*, **29**, 202–206.

Stommel, H. (1958): *Deep Sea Res.*, **5**, 80–82.

Stommel, H. and A. B. Arons (1960): *Deep Sea Res.*, **6**, 140–154.

Stössel, A. and P. Lemke (1990): *J. Geophys. Res.*, **95**, 9539–9555.

Suginohara, N. (1978): *Oceanography as Environmental Science 2*, ed. by S. Horibe, Tokyo University Press, pp. 234–245 (in Japanese).

Takano, K. (1978a): *Oceanography as Environmental Science 2*, ed. by S. Horibe, Tokyo University Press, pp. 27–44 (in Japanese).

Takano, K. (1978b): Numerical Method for the Study of the Ocean, Part I, 79 pp. (unpublished manuscript).

Tokioka, T., M. Yamasaki and N. Satoh (1993): *Kisho no Kyoshitsu 5, Numerical Simulation in Meteorology*, Tokyo University Press (in Japanese).

Yukimoto, S., M. Endoh, Y. Kitamura, A. Kitoh, T. Motoi and T. Tokioka (1996): *Climate Dynamics*, **12**, 667–683.

Chapter 7

Satellite Remote Sensing of the Air–Sea Interaction

Hiroshi KAWAMURA

Ocean–Atmosphere Interactions, Ed. Y. Toba, pp. 239–297.
© by TERRAPUB / Kluwer, 2003.

Satellite Remote Sensing of the Air–Sea Interaction

Hiroshi KAWAMURA

The sea is calm to-night,
The tide is full, the moon lies fair
Upon the straits ...
Come to the window, sweet is the night-air!

—*Matthew Arnold*

7.1 SATELLITE OBSERVATION OF OCEAN AND SEA SURFACE FLUXES

7.1.1 Air–sea interaction and satellite remote sensing

Observation of the earth from a space-borne satellite revolving around the earth is called "Satellite Earth Observation." When the satellite measurements target air–sea interaction that takes place over the global oceans, processes near the sea surface at the bottom of a thick atmosphere need to be observed. Electromagnetic waves of various wavelengths transfer information about the near sea-surface processes to the observation platform at high altitudes. The observable physical quantities depend on the used wavelengths and the employed technology of sensors.

Sea-surface fluxes are defined as an exchange rate of physical quantities per unit area of the sea surface and unit time: these are, momentum flux, energy flux, and material flux. The exchanged energy is heat energy and radiant energy; the exchanged materials are water (vapor, rain) and gases (air, carbon dioxide, etc.). Such sea-surface flux is ubiquitous over the ocean that covers 70 percent of the entire earth surface. Both the sea-state and the atmospheric-state vary as the sea-surface flux varies, and thus, it is a necessity to estimate the flux with sufficient resolution for both atmospheric and oceanographic research.

Satellite remote sensing is a powerful tool for observing the basic processes of air–sea interaction, such as momentum flux, heat flux and water flux over a wide ocean. In order to realize satellite observation of the sea-surface fluxes over a wide ocean, basic techniques of satellite remote sensing and their applications will be discussed in this chapter.

7.1.2 Sea-surface fluxes

The surface wind causes momentum flux at the sea-surface. The ocean current has almost no influence on atmospheric motion since its velocity is much smaller than that of atmospheric motion.

Broadly speaking, there are two types of energy fluxes at the sea-surface: radiation fluxes and turbulent fluxes. Radiation fluxes are shortwave and long-wave radiation. Turbulent fluxes are sensible heat flux and latent heat flux. Latent heat flux is directly related to vapor transportation from the ocean to the atmosphere.

a) Radiation flux at the sea-surface

The ocean receives solar radiation, and emits electromagnetic waves from the surface. The peak energy of solar radiation is at the visible wavelength (0.4–0.7 μm), while radiation from the earth's surface is at the infrared band (3–100 μm), whose wavelengths are much longer than the visible wavelength. Thus, the former is called shortwave radiation, and the latter long-wave radiation. The exchange of energy between space and the earth is made through short wave and long wave radiation.

When the distance between the earth and the sun is at an average and the sun is at normal incidence angle, the received radiant energy at the top of the atmosphere is 1367 W m^{-2} (Yoshino *et al.*, 1985). However, the radiant energy decays before reaching the earth's surface due to atmospheric absorption and scattering. The decay depends on the atmospheric condition: for example, only 60–70% of the radiation reaches the ground in Japan during summer-time.

All objects emit electromagnetic radiation that depends on their temperature. The radiant energy from an object is a function of its temperature and emissivity that depends only on the nature of the substance. According to Kirchhoff's law (see Section 7.3.3), when the emissivity of an object is large, the object effectively absorbs incoming radiation. An object that perfectly absorbs radiant energy at all wavelengths radiates energy most effectively at a given temperature. Such an ideal object is called a "black body." The emissivity of a black body is 1, and the heat radiation from a black body with temperature T is given by the following Planck's formula, including wavelength dependencies:

$$B_\lambda = \frac{2hc^2}{\lambda^5 \cdot \left[\exp\left(\dfrac{hc}{kT\lambda} \right) - 1 \right]}. \tag{7-1}$$

Here, B_λ is the spectral brightness per wavelength band (W sr^{-1} m^{-3}), λ (m) is electromagnetic wavelength, h is Planck's constant (6.63 × 10^{-34} J s), k is Boltzmann constant (1.38 × 10^{-23} J K^{-1}), and c is the speed of light (m s^{-4}). The emissivity is defined as the ratio of the actual radiant energy from a body to the radiant energy from the black body with the same temperature. Spectral irradiance of solar radiation resembles that of a black body at approximately 5900 K. Figure 7-1 displays the spectral irradiance of black bodies with 5900 K and 290 K, which is near average temperature of the earth.

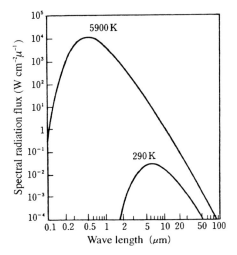

Fig. 7-1. Radiation spectra emitted from black bodies at 5900 K and 290 K.

The ocean temperature ranges between –2°C (freezing) to about 35°C. Thus the electromagnetic radiation has a peak emission at the near-infrared band (9–11 μm). At this wavelength band, the emissivity of the ocean surface is near 1 (0.98–0.99), almost a black body.

The total brightness from the ocean surface can be calculated by integrating Eq. (7-1) over all wavelengths. For example, assuming an ocean temperature of 20°C, the corresponding black-body radiant energy is estimated as 412 W m^{-2} using Stephan–Boltzman's formula, which is an integration of Eq. (7-1),

$$E = \sigma T^4 , \qquad (7\text{-}2)$$

where σ is the Stefan–Boltzmann constant (5.67 × 10^{-8} W m^{-2} K^{-4}). By multiplying 0.98–0.99 of the emissivity of seawater, the radiant energy from the ocean surface is obtained as 404–408 W m^{-2}.

The atmosphere absorbs a part of the long-wave radiation, and emits it. Clouds absorb most of the long-wave radiation and emit it. Ocean surface emits long-wave energy, while at the same time receives long-wave energy from the atmosphere and clouds. Thus, both radiation and absorption needs to be taken into consideration in order to estimate the energy budget at the long-wavelength band at the sea surface.

Figure 7-2 is a schematic of the shortwave and long-wave radiation fluxes at the ocean surface. Satellite remote sensing that uses the visible and infrared electromagnetic band has a close connection to this radiation process between the earth and space (Section 7.2.3).

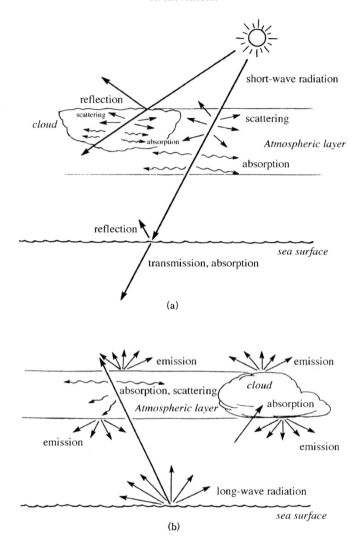

Fig. 7-2. (a) Processes of shortwave radiation in the atmosphere and sea surface, and
(b) long-wave radiation emitted from the surface and processes in the atmosphere.

b) Turbulent fluxes at the sea surface

Momentum, heat energy and material exchange between the ocean and the
atmosphere are caused by turbulence generated by the wind in air and water near
the sea surface. The basics of those were described in Chapter 2. The most
fundamental method to estimate turbulent fluxes is called the eddy-correlation
method. The fluctuating components of wind speed, temperature and relative
humidity are measured and the products of those are averaged over time.

However, the bulk-method is the most commonly used method for estimating turbulent fluxes. The methods require macroscopic physical quantities such as an average wind speed and temperature. The spatially-averaged physical quantities are obtained through satellite observations. For example, an algorithm used for wind speed estimate is designed to give an average wind speed at a height of 10 m. Therefore, estimation of sea surface fluxes using satellite-observed physical quantities is made through usage of a bulk formula that relates the sea surface fluxes to the satellite-observed physical quantities with bulk coefficients.

In general, bulk coefficients for turbulent fluxes are a function of wind speed, temperature and relative humidity at 10 m high. Therefore, when those physical quantities are measured at a different height, turbulent fluxes are estimated by either using a bulk coefficient defined for that height, or by adjusting the observed physical parameters to those for the height of 10 m. For temperature T_{10} (°C), relative humidity q_{10}, and wind speed U_{10} (m s^{-1}) at 10 m high, sea surface wind speed 0 m s^{-1}, sea surface temperature T_s, and sea surface relative humidity q_s, the bulk formula that gives the momentum flux τ (kg m^{-1} s^{-2}), sensible heat flux Q_H (W m^{-2}), and latent heat flux Q_E (W m^{-2}), are given as:

$$\frac{\tau}{\rho_a} = u_*^2 = C_D U_{10}^2 , \qquad (7\text{-}3)$$

$$\frac{Q_H}{\rho_a C_p} = C_H (T_s - T_{10}) U_{10} , \qquad (7\text{-}4)$$

$$\frac{Q_E}{\rho_a L} = C_E (q_s - q_{10}) U_{10} , \qquad (7\text{-}5)$$

similar to Eqs. (2-116)–(2-118). In the formula, u_* is the friction velocity, C_D, C_H, and C_E are the bulk coefficients to estimate momentum, sensible heat and latent heat fluxes, respectively, for the height of 10 m. Various researchers have presented a variety of formulas (for example, see Fig. 2-10). The air density is expressed as ρ_a (1.24 kg m^{-3} for atmospheric pressure, 20°C dry air), C_p (1.006 $\times$ 10^{-3} J kg^{-1} K^{-1}) is the specific heat at constant pressure for moist air, and L is the latent heat of vaporization (2.500 $\times$ 10^6 J kg^{-1}).

Let us estimate the sensible and latent heat fluxes for winter in the Sea of Japan when the exchange is vigorous. Figure 7-3 shows typical cloud formation during the winter in the Sea of Japan when the monsoon wind blows. Except for the windward side near the continental coast, the Sea of Japan is covered with streaky clouds, whose altitude is around 1000–1500 m. The mountains at the backbone of Japan block those clouds, and therefore the Pacific coast is cloud-free. This is the typical weather pattern of Japan in winter.

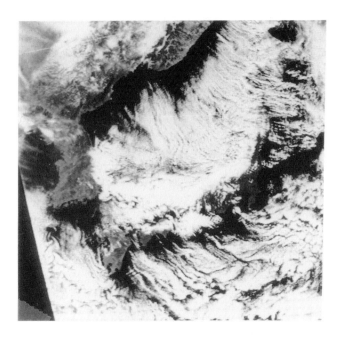

Fig. 7-3. A VHRR visible image around the Sea of Japan obtained on 28 December 1988.

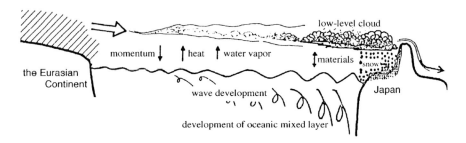

Fig. 7-4. Air–sea interactions under condition of winter monsoon blow.

A schematic illustration of the air–sea interaction that takes place along the wind direction in the Sea of Japan is shown in Fig. 7-4. Cold and dry Siberian Air formed in the Eurasian continent receives large volumes of sensible heat and vapor. Thus the bottom layer becomes unstable, creating a cumulus (Yoshino *et al.*, 1985). This causes heavy snow in the coast of the Sea of Japan.

We can estimate the sensible and the latent heat as follows. Using the December climatological data of Vladivostok, we can approximate the temperature of the cold air as –10°C, relative humidity 60%, and the water temperature at the mid-Sea of Japan as 10°C. With 15 m s^{-1} wind, the latent heat is 426 W m^{-2} and sensible heat is 333 W m^{-2}, using Kondo's formula (Kondo, 1994).

7.1.3 Brief history of satellite oceanography

The most difficult task in oceanography and marine meteorology is to "conduct observation in the ocean." From the early days of oceanography, ships played an important role as the observation platform for oceanographic research. In modern times we have new observation platforms in the form of orbiting satellites from which ocean data is provided at a rate never before experienced by researchers, with potential for even better results in the future. The advantage of satellite ocean observation proved to be its capability of supplying good-quality global ocean data at high sampling rate.

Oceanographic research using satellite remote sensing data became popular in the late 1970s and early 1980s. Seasat (Sea Satellite) was launched in 1978 carrying various microwave sensors. Seasat sensors, such as a microwave scatterometer, microwave altimeter, microwave radiometer, and synthetic aperture radar have been improved and have become indispensable tools for satellite oceanography.

The Advanced Very High Resolution Radiometer (AVHRR: visible and infrared radiometer) launched by NOAA series satellite has also been in operation since 1978, providing global sea surface temperature. A number of geostationary meteorological satellites were put into orbit over the equator in the 1970s. Satellite-observed data started to become available for scientists in the early 1980s. By the late 1980s, fast and reliable observation of the ocean and the marine atmosphere by satellite-borne sensors was realized and researchers were increasingly able to utilize this satellite-generated data.

In addition to remote sensing using visible and infrared wavelengths, microwave remote sensing was introduced in the 1980s. By the 1990s, a variety of satellite-borne sensors were in use providing information about a variety of phenomena on the earth. Those are summarized in Table 7-1.

Drastic development also occurred in computers and networks. Circuit miniaturization saw great progress and, by the end of the 1980s, cheap work stations became the main tool for satellite data analysis and processing. Advances in computer network technology accelerated the distribution of this data. The database that provides daily satellite images of Japan was established as a result of these technological advances (Kawamura, 1995a).

Only two decades have passed since the satellite observational data began to be used for air–sea interaction research, and monitoring skills are still developing. It will take some time until the measurement technology and the methodology are established and satellite ocean observation becomes a general tool for air–sea interaction research. The author estimates that accumulation of know-how to utilize satellite measurement for ocean observation will continue through the early years of the 21st century, after which satellite ocean monitoring will be established as an integral part of the earth environment observation system.

Table 7-1. Short history of satellite ocean observation.

Year	Sensor/Satellite
1977	VISSR/GNS1 (Start of GMS series satellite)
1978	SASS•SMMR•ALT/Seasat, SMMR/Nimbus 7
1979	AVHRR/TIROS-N (Start of NOAA series satellite)
1985	ALT/Geosat
1987	VTIR•MSR/MOS1, SSMI/DMSP
1991	AMI•ALT•ATSR/ERS-1
1992	ALT/TopexPoseidon

This table shows the launch years of satellites carrying sensors, which are visible-infrared radiometer, microwave radiometers, microwave scatterometers and altimeters. These snsors are described in this text because they are closely related to the study of air–sea interaction.

7.1.4 The advantage of satellite ocean observation

Satellite remote sensing is indirect measurement of oceanic parameters, while previous observations using platforms in the seas were direct measurements. Acoustic ocean observation is a remote measurement, but it requires a platform to stay in contact with the sea. Indeed the conventional measurements of the ocean parameters are called "*in situ*" observation. The use of orbiting satellites as remote platforms was quite new and unique.

Since the beginning of the ocean research, the main ocean observational platforms have been ships. Since this was the basic source of information about oceans, the Brussels International Oceanography and Meteorology Conference agreed to request ships voyaging in the world ocean to make regular observations (Uda, 1978) and report them in a consistent manner. After that, the ocean meteorological parameters, such as sea surface temperature, air temperature, wind speed and direction, and clouds were well reported from the ships and used for various research applications. Because of this well-organized observation system, the number of ship-reported meteorological parameters is greater than that obtained from other ocean observations.

Figure 7-5 displays the ship-reported observations during 1 July to 9 October in 1978. Most points are from merchant ships and correspond to major voyage routes. The observation points are concentrated in the Northern Hemisphere, and are scarce in the Southern Hemisphere, especially around the Antarctica. The observational period of three months corresponds to that of the Seasat operation, which was the first satellite to carry microwave sensors to observe the ocean. Figure 7-6 shows the wind vectors observed by Seasat-SASS during 14–15 September 1978. Wind speeds are shown in the observational swaths of SASS (Seasat-A Sacatterometer System) and the schematic streamlines by arrows. SASS had 50-km spatial resolution and observed ocean wind vectors within the two 600 km swaths along the satellite track. The SASS

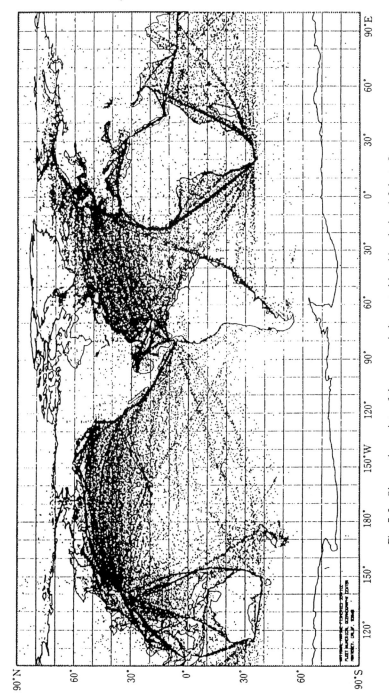

Fig. 7-5. Observation points of the ocean wind reported by ships during 1 July to 9 October 1978 (Freilich, 1985). The observation points are gathered from the main ship routes.

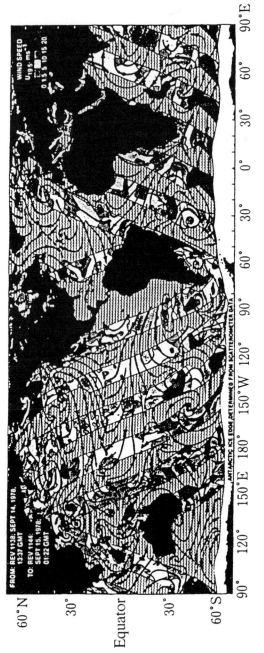

Fig. 7-6. The distribution of surface winds over the global oceans based on observations by SASS on board Seasat during 14–15 September 1978 (Freilich, 1985). The wind speeds are indicated by degree of shading in the satellite swaths, and there is no observation in the regions with wavy patterns. The schematic stream patterns are illustrated by arrows, and the atmospheric fronts dashed lines.

observed about 15 million points during the observation period. The observation point shown in Fig. 7-5 is 0.119 million. Thus the satellite observed about 120 times more than the ship reports during the same period.

Measurements of ocean surface winds have been one of the most important tasks of ship reports. The surface wind is the essential parameter to estimate momentum and heat fluxes between the air and the sea. However, since surface winds are rapidly changing physical quantities, it is difficult for ship reports to observe them in the world ocean in detail. Below, we list advantages of the microwave scatterometer, which was specially developed to measure wind speed and direction:

(1) High-spatial resolution enabling the synoptic scale description and wide coverage.
(2) Measurements regardless of the ocean weather, night and day.
(3) Regular global observations.
(4) Measurements equally conducted in both space and time.
(5) Uniform quality of measured wind vectors.

7.2 ATMOSPHERE AND THE RADIATION FLUXES

7.2.1 Satellite remote sensing and the atmospheric window

Observations of distribution and change of sea-surface fluxes are essential for air–sea interaction research, as described in the previous sections. In this section, the principles of the satellite remote sensing technique will be given with an emphasis on measurements of the physical quantities necessary for estimating sea surface fluxes. The ocean is observed by satellites, extracting the information contained in the electromagnetic waves of various wavelengths that were transmitted from the ocean surface.

The electromagnetic waves are absorbed and scattered by gas molecules and aerosol in the atmosphere. Thus, the electromagnetic waves used for satellite observation must be in the bands where the atmospheric transmittance is large. The bands with high transmittance are called the "atmospheric windows." Figure 7-7 shows the transmittance for different wavelengths. It shows that the visible, infrared, and microwave electromagnetic wavelengths have high transmittance. These wavelength bands also contain information on parameters related to the air–sea interaction.

7.2.2 Passive and active sensors

Satellite remote sensing is classified into two: the passive sensor measures various physical quantities by receiving electromagnetic waves radiated from the ocean; the active sensor receives electromagnetic waves transmitted from the satellite and reflected back from the sea surface. The passive sensor is commonly called a radiometer. There are visible radiometers, infrared radiometers, and microwave radiometers (Section 7.1.3).

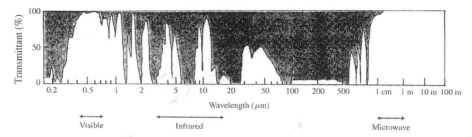

Fig. 7-7. Transmittance of electromagnetic waves for remote sensing of the ocean.

The majority of active sensors for observation of the sea-surface fluxes employ radar technology; they are the microwave scatterometer, microwave altimeter, and synthetic aperture radar. These sensors give information on sea-surface roughness (Section 7.4) and can retrieve physical parameters for estimation of the turbulent fluxes.

Using a radar antenna small enough for satellite equipment, the synthetic-aperture radar (SAR) synthesizes a large aperture antenna in order to achieve high spatial resolution. It images the sea-surface roughness generated by the wind and the turbulence in water. Its spatial resolution is high (order of 10 m) and the observational coverage is small. Since the observational frequency is poor for a fixed point, the SAR is not suitable for research on the air–sea interaction discussed here.

7.2.3 Visible, infrared radiometer and radiation fluxes

The passive sensors receive oceanic information through the electromagnetic waves radiated from the sea surface and transmitted to the satellite through the atmosphere. As mentioned in Section 7.1.2(a), all objects radiate electromagnetic waves as a function of temperature. The sun and the earth are no exception. Energy balance of the earth is achieved by a thermal equilibrium of incident solar energy and radiating infrared waves to outer space. Visible and infrared radiometers utilize a part of this radiation mechanism.

Four visible light images of the globe obtained for one day by the visible radiometer aboard the geostationary meteorological satellite, "Himawari" are presented in Fig. 7-8. The images from "Himawari", changes through the dawn, daytime, and dusk. The bright white parts are clouds, and the lands and the ocean regions are darker. This is due to the difference in their albedo for the visible light; albedo is the ratio of the reflecting energy to the incoming energy of the sunlight reflected from an object (Section 7.3.3). Typical values of the albedo are 0.03–0.1 in the ocean, 0.1 in the forest, and 0.20–0.45 in the desert. The cloud albedo varies from 0.2 to 0.8, depending on its conditions and is much larger than the ocean albedo. The visible radiometer receives sunlight reflected from the earth.

Fig. 7-8. Visible images of the globe taken by a visible radiometer aboard the geostationary meteorological satellite, "Himawari", on 23 July 1987. (a) 8:00, (b)12:00, (c)15:00 and (d)18:00.

Clouds, sea and lands in the visible image are distinguished by the difference of received radiation energy representing their albedo. The earth is imaged with the sun acting as a light source. Therefore, at night the earth is invisible. Since these images were taken during wintertime of the Southern Hemisphere, the Antarctic region is always dark due to the polar night.

Figure 7-9 is the infrared image from "Himawari." The temperature is low in the whiter part. The clouds in the high altitudes and the polar areas are white because of their low temperature. Here temperature differences correspond to the earth components. There are no shaded regions as in the visible image since the temperature-dependent radiation from the earth surface is imaged. Therefore, the infrared observation can be made without any lighting source. "Himawari" can present images of the clouds even during nighttime.

Fig. 7-9. Infrared image of the globe by "Himawari" at 15:00 on 23 July 1987. This image was taken together with the image of Fig. 7-8(c). Note that the shaded region in the visible image due to nighttime is observable in this infrared image.

7.2.4 Observation of sea-surface solar radiation by VISSR aboard geostationary meteorological satellite

In the early 1970s, in order to monitor the distribution and motion of clouds, a combination of the satellites carrying visible and infrared sensors was placed into geostationary and polar orbits. Since then, a number of geostationary and polar orbit satellites have been used on a regular basis for global meteorological and oceanographic observations. These satellites are replaced regularly in order to achieve continuous observations and monitoring. NOAA satellites, a representative polar orbit satellite, have a visible and infrared radiometer, which is called the Advanced Very High Resolution Radiometer (AVHRR). Geostationary Meteorological Satellites (GMS) in operation are two from the USA, and one each from Japan, India, and the European Community. The Japanese satellite

"Himawari" has a Visible Infrared Spin-Sea Radiometer (VISSR) on board and observes half the globe from the equatorial orbit. The altitude of the GMS satellite is 36000 km in order to have a geostationary orbit. The VISSR has one visible channel and one infrared channel (Kimura and Murayama, 1990).

The advantage of the GMS is its ability to observe the earth from a stationary observation point relative to the revolving earth. In other words, a GMS can continuously observe a phenomenon at a particular location. Currently, a VISSR monitors every hour, tracing cloud motions. The spatial resolution of the infrared channel is about 5 km square at the equator. The visible channel (0.5–0.75 μm) has four detectors with 1.25 km spatial resolution. However, in order to compensate for the sensitivity difference of each sensor, adjacent 4×4 pixels are averaged (see for detail JMA, 1984; Sasaki et al., 1990).

The solar radiation at the earth's surface can be estimated from the densely sampled visible images of the GMS (Gautier et al., 1980). In the clear sky, the atmospheric absorption and scattering of solar radiation are parameterized; in the cloudy sky, reflection and absorption of solar radiation due to clouds are added. Daily averaged solar radiation from the VISSR/GMS is given by Kizu et al. (1990). Global distribution of solar radiation is estimated using the cloud distribution data (ISCCP: International Satellite Cloud Climatology Project; Rossow and Schiffer, 1991) compiled from observations of the global geostationary satellite and the polar orbit satellite (Rashke et al., 1987; Bishop and Rossow, 1991).

7.3 SEA SURFACE TEMPERATURE MEASUREMENT BY SATELLITE

7.3.1 Introduction

SST (Sea Surface Temperature) is one of the most fundamental parameters for air–sea interaction research. SST is not just the boundary conditions for oceanic and atmospheric motions but also an indicator of the dynamical interaction between the atmosphere and the ocean. In order to estimate air–sea fluxes, the absolute value of the SST with sufficient accuracy is needed.

Since satellite infrared remote sensing came into practice, mostly using the NOAA AVHRR infrared observational data, extensive research has been conducted on how to obtain the absolute value of SST. Now, with the atmospheric correction technique, SST observation by the AVHRR is in practical use (NOAA, 1991).

7.3.2 NOAA satellite and the AVHRR sensor

a) NOAA satellite series

The GMS together with the NOAA polar-orbiting satellite (USA) covers the whole earth surface; the former is positioning around the equator and the latter flies over the both poles, neither of which can not be seen from the geostational orbit. The height of the NOAA orbit is about 850 km and the inclination of the orbital plane to the earth's equatorial plane is 98 degrees. Two vehicles are in operation at a time. One orbits from the north pole to the south pole (descending orbit), and the other orbits from the south pole to the north pole (ascending orbit). One satellite flies twice over the same area of the earth in a day. Table 7-2 summarizes the operational periods and the sensors on board the NOAA series satellite.

AVHRR sensors on board the NOAA satellite series have been observing the whole globe for more than a decade, while being a representative visible and infrared radiometer. The observation data is stored on a data recorder on board the satellite, and transmitted to the ground receiving station of the NOAA when it passes over the United States.

One of the characteristics of the NOAA series satellites are their capability to transmit observational data to the ground in real time. Because of this system, local receiving stations have been established in many countries, not only in the United States, and local user communities have been formed. The NOAA satellites have two data broadcasting methods; the first is APT (Auto Picture Transmission) in which a low-resolution image of AVHRR data with simple corrections is sent by a FAX signal. The second is HRPT (High Resolution Picture Transmission) in which full digital observation data of most sensors is transmitted digitally to the local ground stations. Such data can be received when the satellite flies over the local station though the observation data only covers the area around the station. Figure 7-10 illustrates the observable area of the AVHRR receiving station at Tohoku University in Sendai, Japan.

b) AVHRR sensor and data processing

The AVHRR sensor scans its detector ±55.4 degrees transversely to the direction of the satellite pass and observes the visible and the infrared radiation strength from the earth surface and the clouds. One receiving station can cover the HPRT data for about 3000 km wide (2048 pixels) and about 5000

Table 7-2. NOAA satellites and AVHRR sensors.

NOAA satellite	Operation period	AVHRR sensor
TIROS N	1978–1980	AVHRR/1
NOAA 6	1979–1983	AVHRR/1
NOAA 7	1981–1985	AVHRR/2
NOAA 8	1983–1985	AVHRR/1
NOAA 9	1985–	AVHRR/2
NOAA 10	1986–	AVHRR/1
NOAA 11	1988–1994	AVHRR/2
NOAA 12	1991–	AVHRR/2
NOAA 13	(1993)	AVHRR/2
NOAA 14	1995–	AVHRR/2

NOAA-13 did not function though launched successfully. AVHRR/1 has four channels, and AVHRR/5 five channels.

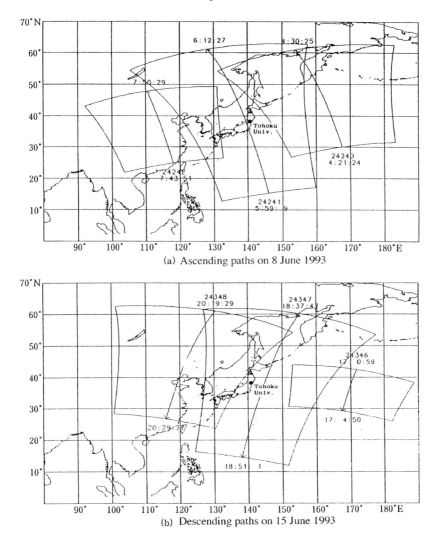

(a) Ascending paths on 8 June 1993

(b) Descending paths on 15 June 1993

Fig. 7-10. Coverage of the AVHRR scene in the HRPT data received at a local station in Sendai, Japan (Kawamura *et al.*, 1993a). (a) Ascending passes on 8 June, 1993, and (b) descending passes on 15 June, 1993. At the beginning and the end of the passes in the coverage, the start time and the end time of reception are indicated. The orbit number is also shown at the beginning of the passes.

km (4500 lines) in the direction of the satellite pass (Figs. 7-10 and 7-11). The spatial resolution of the AVHRR sensor is about 1.1 km at nadir viewing, but as the distance becomes larger from the image center, the resolution reduces to 6.5 km.

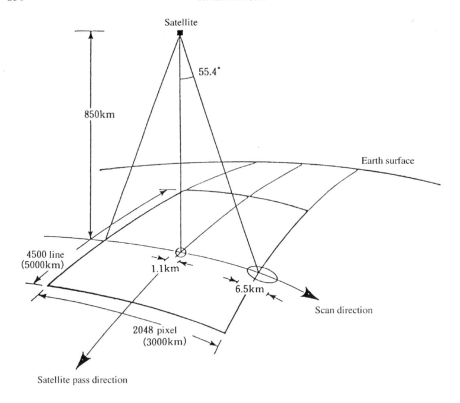

Fig. 7-11. A schematic of geometry in terms of AVHRR observation of the earth surface.

The first and the second channels of AVHRR are visible and near-infrared wavelength bands, respectively. Since both measure the sunlight reflection from the ground and from clouds, they are sometimes called the solar channels. The third, the fourth and the fifth channels for SST measurement uses infrared wavelengths and measure the heat radiation from the earth's surface. Some AVHRRs have the fifth channel, but some do not (Table 7-2). The global SST data sets are generated using the infrared data of the AVHRR with the fifth channel.

Figure 7-12 displays the AVHRR images from the HRPT data received at the center orbit shown in Fig. 7-10(a). Hereafter, we call this the original data. The channel 2 and 4 images are shown in the left and right of Fig. 7-12, respectively. Japan is seen at the center of the images, which are mostly covered by clouds except for a part of the Tohoku area. Part of Russia is seen in the upper left corner.

The original data shown in Fig. 7-12 is in counts (0–1024 integers) representing the radiation strength measured by the sensor. The digital data, which cannot be used for research yet, requires further post-processing. The original AVHRR images (Fig. 7-12) are quite distorted due to differences in the spatial resolution from the center to the edge of the image (Fig. 7-11) and

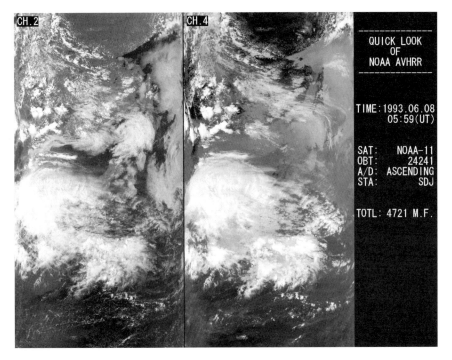

Fig. 7-12. AVHRR images of the whole coverage corresponding to the central pass (No. 24241) shown in Fig. 7-10(a) (Kawamura *et al.*, 1993a).

due to the earth rotation during the observation period. For research purpose, such distortion needs to be corrected and be projected on a map. These corrections are called geometrical transformation.

The digital counts from each AVHRR channel are transformed to physical quantities using the calibration data of each detector. The first and the second channels are normalized such that the solar radiation strength scattered from a Lanbertian at nadir becomes 100. This normalized value is called albedo, which does not correspond to the original definition that is customarily used for satellite observation.

Counts from the infrared channels are transformed to temperature (brightness temperature). Since the characteristics of the infrared detectors drift with time, they are calibrated both by the temperature of an internal reference black body and outer space. The calibrated counts are transformed to radiation strength and then the brightness temperature T is computed using Eq. (7-1), assuming that the radiation is from a black body with temperature T. This processing is called brightness temperature transformation. The bright temperature contains the atmospheric influence, so it does not correspond to the exact temperature of the sea surface.

Fig. 7-13. Image of the brightness temperature around Japan produced by AVHRR data through geometric and radiometric transformations. The image was produced by compositing five AVHRR brightness temperature images obtained around 20 April 1993.

Figure 7-13 shows the image of SST distribution obtained from the original data through geometrical transformation and brightness temperature transformation. The image of a Mercator projection is created from several scenes of brightness temperature distribution received over a few days. The clouds are eliminated sampling the highest temperature at each pixel. The brightness temperature is lower than the actual SST due to the attenuation of infrared radiation emitted from the sea surface in the atmosphere. However, the SST patterns indicating the oceanic features are useful because relative values of temperature are mostly kept. It is useful for detecting a large SST gradient front; the Kuroshio Current south of Japan is seen as a dark band, which is called the warm core and meanders near the coast of Kii peninsula.

7.3.3 Sea surface, sea surface temperature and infrared radiation

The sea surface plays a significant role in remote sensing of the ocean and the marine atmosphere. The electromagnetic characteristics of air and seawater are quite different. Electromagnetic waves can propagate more freely in the air than in seawater, which strongly absorbs the electromagnetic waves. Therefore, the sea surface forms a boundary for the electromagnetic waves.

a) Sea surface and the electromagnetic wave

Let $E_i(\lambda)$ be the electromagnetic wave energy incident to a material; $E_r(\lambda)$ the reflected energy, $E_a(\lambda)$ the energy absorbed in the material, and $E_\tau(\lambda)$ the transmitted energy. λ is the electromagnetic wave length. The following relation holds:

$$E_i = E_r(\lambda) + E_a(\lambda) + E_\tau(\lambda);\qquad (7\text{-}6)$$

Or,

$$1 = \frac{E_r(\lambda)}{E_i(\lambda)} + \frac{E_a(\lambda)}{E_i(\lambda)} + \frac{E_\tau(\lambda)}{E_i(\lambda)}$$
$$= r(\lambda) + a(\lambda) + \tau(\lambda),\qquad (7\text{-}7)$$

where $r(\lambda) = E_r(\lambda)/E_i(\lambda)$ is the reflection coefficient, $a(\lambda) = E_a(\lambda)/E_i(\lambda)$ the absorption coefficient, and $\tau(\lambda) = E_\tau(\lambda)/E_i(\lambda)$ the transmission coefficient. The last coefficient for the whole atmospheric layer was discussed earlier in Section 7.2.1.

According to Kirchhoff's law, materials that absorb electromagnetic waves well with a certain wavelength transmit it well with the same wavelength. In other words, for an object at thermal equilibrium, the following equation holds;

$$e(\lambda) = a(\lambda).\qquad (7\text{-}8)$$

Here $e(\lambda)$ is the emissivity. The black body, which causes the most efficient emission of thermal radiation, is the best absorber. Thus, for a black body, $e(\lambda) = a(\lambda) = 1$.

Let us consider behavior of the visible, infrared, and microwave electromagnetic waves near the sea surface, which is formed by pure seawater.

Table 7-3. Reflection coefficients at the sea surface and skin depth for the visible, infrared and microwave wavelengths.

Electromagnetic wave	Wavelength	Reflectance	Skin depth
Visible band	0.4 μm	0.01–0.02	75 m
Infrared band	10 μm	0.01–0.02	3 μm
Microwave band (5 GHz)	1.5 cm	0.64	0.5 cm

The reflection coefficient $r(\lambda)$ for still seawater and the skin depth, at which electromagnetic waves can penetrate into the seawater, are shown in Table 7-3. For both visible and infrared electromagnetic waves, the ocean is a good absorber. The visible light has a small reflection coefficient at the surface; it mostly penetrates into the ocean and is absorbed in the seawater. Seawater deeper than 75 m absorbs most of the visible electromagnetic waves. The absorbed electromagnetic waves are stored as heat and raise water temperature.

On the other hand, infrared electromagnetic waves are absorbed in a thin layer at the surface. They are completely absorbed within a 0.02-mm layer. Thermal radiation at the infrared wavelengths from the ocean comes from this thin layer. The temperature of this thin layer is called skin temperature. Since the temperature can be considered uniform within this thin layer, SST is equivalent to skin temperature. The emissivity of seawater is near that of the black body (0.98–0.99). Thus, SST can be principally detected by thermal radiation energy from the sea surface.

Microwaves are absorbed within a thin layer as well. However, microwaves have a high reflectance at the sea surface. Therefore, the sea surface is not the black body for microwaves. Interaction of microwaves and the sea surface will be described in detail in Section 7.4.

b) Sea surface fluxes and the sea surface heat radiation

SST obtained by ships and buoys is called the bulk temperature or the *in situ* SST. The temperature of seawater, sampled by drawn buckets and through the intake for cooling of the ship engine, is measured by thermometers. It is considered that the *in situ* SST corresponds to the water temperature at depths of a few tens of centimeters to ten meters. Researches on long-term air–sea interaction are based on the *in situ* SSTs.

An infrared radiometer indirectly measures temperature of the 0.02-mm thin layer at the sea surface. It is nearly impossible to measure directly the temperature of such a thin layer. Furthermore, it is impossible to conduct such measurement in the ocean.

There can be a difference between the remotely sensed sea surface temperature (T_s) and the directly measured bulk temperature (T_b). Let us define the difference; $\Delta T = T_s - T_b$, which varies depending on heat and momentum fluxes and associated turbulence (Saunders, 1967; Katsaros, 1980). As the wind speed increases, the sea roughens. Because of the wind waves and turbulence in the surface layer, the measurement of ΔT becomes difficult. Up to now, it has been considered $|\Delta T| < 1°C$ in the outer ocean (Stewart, 1985). Figure 7-14 is a schematic diagram describing the difference of skin and bulk temperatures and the sea surface heat flux (Stewart, 1985).

When the solar radiation is strong, the ocean is heated due to absorption of light in the seawater. The absorption is large near the surface, creating a large gradient of temperature in the vertical direction $(\Delta T > 0°C)$. When a warm and humid wind blows, the sea surface is also heated. In such a case, the skin temperature becomes higher than the bulk temperature. It is dynamically stable

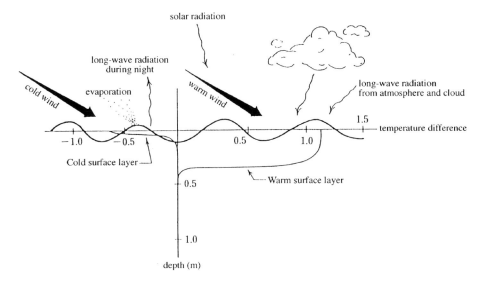

Fig. 7-14. A schematic indicating relations among the skin and bulk sea surface temperatures and the surface heat fluxes (Stewart, 1985).

when the temperature in the upper layer is higher, and the stable state is kept until turbulent mixing or surface cooling destroys the temperature gradient.

When infrared radiation from the sea surface dominates at night under weak wind and clear sky conditions, SST lowers and the skin temperature becomes lower than the bulk temperature. Also, when dry and cold air blows over the ocean, the sea surface is cooled and ΔT can become less than zero. It is dynamically unstable in the case that the surface temperature becomes lower due to heat flux from the sea to the air. When the heat flux forcing surface cooling diminishes, the temperature gradient is broken due to thermal convection. Therefore, a state with negative ΔT is transitional. When the sea surface is cooled down due to infrared radiation without generating turbulence, there can be a stable condition with $\Delta T < 0$ until Rayleigh's critical condition is realized and instability occurs.

7.3.4 Sea surface temperature estimate by infrared radiometer

a) Atmospheric correction

Generally speaking, the brightness temperature measured by satellite infrared radiometer is smaller than the *in situ* measurement. The electromagnetic waves emitted from the sea are absorbed by the vapor in the atmosphere and scattered by aerosol. As shown in Fig. 7-7, there are several infrared bands with high atmospheric transmittance, such as 3.8-μm, 8.5-μm, and 11-μm bands. AVHRR sensors uses 3.7-μm and 11-μm bands. The 3.7-μm band is suitable for the SST measurement since it has higher sensitivity to temperature changes and has a high atmospheric transmission coefficient. However, since the solar light reflection from the earth surface contaminates the emitted signals, it is not suitable for daylight measurement. This band is utilized for measurement of regions with little sunlight reflection and at night.

The most common band used in the SST measurement is the 11 μm band. Since this band, compared to the 3.7 μm band, is strongly influenced by atmospheric vapor, corrections are necessary to obtain SST values. It is called atmospheric correction to remove the effects of atmosphere for retrieval of physical parameters of the earth surface.

Two infrared channels are used in order to correct for attenuation of infrared radiation from the sea surface due to atmospheric vapor absorption. The difference in the absorption characteristics of two infrared wavelengths provides information about the column water vapor. Global SSTs have been monitored using this method from the AVHRR data of NOAA satellites.

The five-channels AVHRR has one 3.7-μm band sensor as the third channel, and two 11-μm band sensors as the fourth and the fifth channel. The method that utilizes the third channel and one of the 11-μm band channels is called Dual-Window method. The method utilizing the two channels of 11-μm band is called Split-Window method. The method utilizing all three channels is called Triple-Window method. The Dual-Window method and the Triple-Window method are used during the nighttime since they are contaminated by sunlight reflection.

b) *Examples of the Sea Surface Temperature estimates*

The brightness temperature (T_4) from the fourth channel of AVHRR/NOAA 11 and the bulk temperature (T_b) from the ocean observation buoy at 1 m depth is compared in Fig. 7-15. The ocean observational buoy operated by the Japan

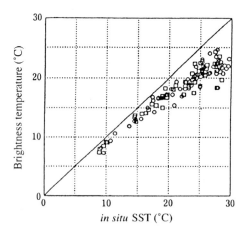

Fig. 7-15. Comparison between the brightness temperature of AVHRR 4th channel and the bulk temperature at 1 m depth (Sakaida and Kawamura, 1992).

Meteorological Agency obtains 11 types of marine meteorological data including the bulk temperature every three hours. The bulk temperature is interpolated in time to match the satellite observation time for comparison. A set of data consisting of the brightness temperatures from the AVHRR at the buoy location and the corresponding bulk temperature was created. Such a data set is called a match-up data set. The *in situ* data (in this case the sea surface temperature from buoys, T_b), which is the target of satellite estimation, is called Sea Truth.

In Fig. 7-15, it is apparent that T_4 is lower than T_b and the tendency becomes larger with increase of the bulk temperature. At $T_b = 30°C$, the difference is about 8°C. This is mainly due to the influence of the atmospheric vapor (Section 7.3.4(a)).

Among the various SST estimation algorithms for AVHRR sensors, the Multi-Channel SST (MCSST) method is most widely used:

$$SST = T_i + \gamma\left(T_i - T_j\right). \qquad (7\text{-}9)$$

The second term of Eq. (7-9) represents the atmospheric vapor correction. Here, T_i and T_j are the brightness temperatures from two wavelength bands, and γ is a constant (Anding and Kauth, 1970). Many modified formulas were proposed later on, and the following formula is presented now including the effect of the satellite zenith angle, ϕ:

$$MCSST = aT_4 + b\left(T_4 - T_5\right) + c\left(T_4 - T_5\right)(\sec\phi - 1) - d. \qquad (7\text{-}10)$$

The third term is a correction for the error due to the path length of the electromagnetic wave propagation, changing with the scan angle (see Fig. 7-11). Constants, a, b, c, and d are empirically determined. The NOAA determined the value of these constants by comparing the AVHRR estimates with the observational SSTs from buoys drifting in the world oceans. This method to determine the SST-derivation coefficients means that "the SST estimated from AVHRR data is tuned against the bulk temperature measured by the drifting buoys."

The formula used for the AVHRR on board NOAA 11 (NOAA, 1991) is

Day: $MCSST = 1.01345T_4 + 2.659762\left(T_4 - T_5\right)$
$$+ 0.526548\left(T_4 - T_5\right)(\sec\phi - 1) - 4.592, \qquad (7\text{-}11)$$

Night: $MCSST = 1.052T_4 + 2.397089\left(T_4 - T_5\right)$
$$+ 0.959766\left(T_4 - T_5\right)(\sec\phi - 1) - 15.520474. \qquad (7\text{-}12)$$

It is known that different formulas should be used for the daytime and the

nighttime estimates of SST. The resultant SSTs through the atmospheric
correction for the data in Fig. 7-15 is presented in Fig. 7-16.

In order to evaluate the satellite-derived parameters quantitatively, we
calculate the Root Mean Square Difference (RMSD) between the satellite and sea
truth parameters. For evaluating the satellite SSTs (MCSSTs) for the bulk SST,
the following formula is used to obtain the RMSD:

$$\mathrm{RMSD} = \sqrt{\frac{1}{N} \cdot \sum_{i=1}^{N} \left(\mathrm{MCSST} - T_\mathrm{b}\right)^2} \, . \qquad (7\text{-}13)$$

In the case of Fig. 7-16, the RMSD is 0.59 degrees.

c) Satellite observation and its validation

Since satellite observation is indirect measurement, oceanic parameter
retrieval from remote sensing measurements is always exposed to errors due to
unexpected changes in the sensor sensitivity and in the atmospheric condition.
One example is volcano eruption; because of the aerosol in the stratosphere,
the SST estimation formula becomes useless. It is difficult to quantify such an
effect and produce the retrieval algorithm beforehand.

The NOAA regularly checks the observational accuracy of satellite-derived
SSTs using the match-up data set including the SSTs from drifting buoys. To
check the accuracy of a satellite estimate through comparison with the sea
truth data is called "validation". An example of validation for satellite-derived
SST is shown in Fig. 7-16. As a measure for validation, the RMSD calculated
from a statistically significant number of the match-up data is used (Section

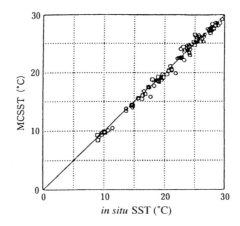

Fig. 7-16. Comparison between the atmospherically-corrected AVHRR-derived SST
and the bulk temperature at 1 m depth (Sakaida and Kawamura, 1992).

7.3.4(b)). The validation indicates that the remotely sensed estimates should lie within the true value ± RMSD. In order to use satellite-derived ocean parameters as a research tool, it is desirable to conduct their validation continuously for a long term and in a wide area.

7.4 MICROWAVE OCEAN OBSERVATION AND PHYSICAL PROCESSES AT THE SEA SURFACE

7.4.1 Physical processes of the sea surface

In Sections 7.2 and 7.3 where remote sensing using the visible and infrared radiometer is described, the sea surface geometry is not discussed in terms of effects on electromagnetic waves. One of the reasons is that the wavelengths of visible and infrared lights is much shorter than the scale of the sea surface undulations. However, the fine structure of the sea surface geometry becomes very important for remote sensing of surface waves and winds using microwaves. Below, the physical processes responsible for the formation of the undulation of the sea surface will be described first (Section 7.4.2), and the principle of microwave remote sensing based on the microwave and sea surface interaction will be presented (Section 7.4.3).

a) Configuration of the sea surface

When wind blows over the ocean, wind waves are generated and start to develop. When a sudden gust blows over still water or over a swell with smooth surface, the sea surface roughens with a dark wavy spot called the cat's paw. It is an important skill in navigating a yacht to recognize such spots when the wind is weak (Kinsman, 1965). These centimeter-scale waves that are generated instantaneously on a still water surface are called initial wavelets. They respond to gustiness of the wind and create wrinkles on the sea surface (Fig. 7-17). These small roughnesses on the water surface are a good microwave scatterer (Section 7.4.3).

The initial wavelets disappear in a short while and become larger wind waves if wind continues to blow. The wind waves grow with time, increasing the wavelength and the wave height. Complicated configurations of the water surface are formed by the wind waves. Let the undulation of water surface be $\eta(x, y; t)$. In order to incorporate the geometrical properties of the sea surface into the microwave back scattering model, $\eta(x, y; t)$ needs to be expressed in a mathematical formulation. Often, a probability density function of the surface slope and spectral energy density of the surface elevation are used. The former is used for models for microwaves incident vertically on the water surface (for example, specular reflection model, Barrick, 1968). The latter is used for the Bragg scattering model in which the microwaves are incident on the surface with large incidence angle (Wright 1968).

Many researches on the probability density function of water surface elevation and the water surface slope have been conducted from the 1950s to the 1960s (Kinsman, 1965). Most of them were based on time series of the elevation

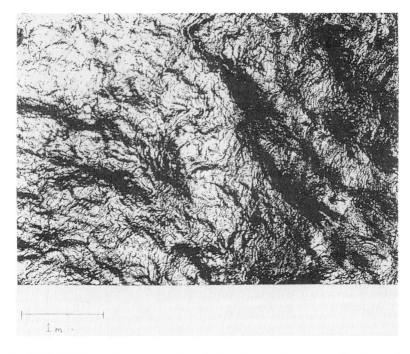

Fig. 7-17. Initial wavelets generated by wind gustiness on the wavy surface in a sun-glitter area.

measurement at a single point, $\eta(t)$. Cox and Munk (1954a) have analyzed aerial photographs of sun glitter at the sea surface and found that the slope distribution is near Gaussian (Fig. 7-18). This result shows that the slope distribution is symmetric in the direction perpendicular to the wind, but is slightly skewed in the wind direction. This is considered to be due to non-uniformity in the stress distribution. Furthermore, there is a linear relation between the wind speed and the root mean square of the surface slope (Cox and Munk, 1954b). This spatial distribution of the surface slope in a wide area have become an important contribution to later works on electromagnetic wave–sea surface interaction and the remote sensing of the ocean.

b) Frequency spectrum of the sea surface elevation
    In microwave backscattering at a large incidence angle, water waves with wavelength of a few centimeters to a few tens of centimeters have an important role. The restoring force of free surface waves with these wavelengths is both gravity and surface tension. The dispersion relation is

$$\sigma^2 = gk + S'k^3 = gk\left(1 + \frac{S'k^2}{g}\right), \qquad (7\text{-}14)$$

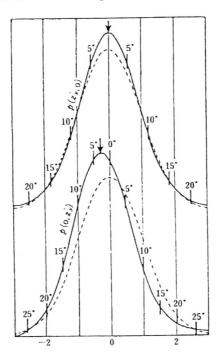

Fig. 7-18. Probability density of the surface slope measured by the photograph of sun-glitter area (Cox and Munk, 1954a). The solid line shows the measured distribution, and the dashed line Gaussian distribution which has the same standard deviation as the measurements. The upper distributions are for the direction perpendicular to the wind, and the lower for the wind direction.

where $S' = S/\rho$ and S is the surface tension between seawater and air, σ the angular frequency, k the wavelength, g the acceleration of gravity, and ρ is the seawater density. The dispersion relation is indicated in Fig. 7-19. The wave with 1.72-cm wavelength has minimum phase speed and the corresponding frequency is 13.5 Hz. Surface gravity waves with a few centimeter wavelengths are called gravity-capillary waves. When the influence of the surface tension becomes smaller ($S'k^2/g$ becomes smaller), they are called short gravity waves (see Fig. 7-19).

Frequency spectra of wind waves plotted in a logarithmic diagram show that wave energy decreases rapidly with the increase of frequency in the high-frequency range corresponding to wavelengths of a few centimeters (Fig. 7-20). Phillips (1958) presented from dimensional consideration a formula to describe the wave spectrum in the high-frequency range:

$$\phi(\sigma) = \beta g^2 \sigma^{-5} , \qquad\qquad (7-15)$$

where β is a constant.

Toba (1973) presented a spectral form:

$$\phi(\sigma) = \alpha_s g_* u_* \sigma^{-4} \qquad (2\text{-}106)$$

$$g_* = g\left(1 + \frac{S'k^2}{g}\right) \qquad (2\text{-}107) \qquad\qquad (7\text{-}16)$$

This formula contains explicitly the friction velocity, indicating an increase of high-frequency energy level with wind speed. According to the microwave backscattering theory in Section 7.4.4, the backscattered energy is proportional to the spectral energy of surface waves at frequencies corresponding to wavelengths of a few centimeters, i.e., at around 10 Hz. Therefore, the spectral representation containing wind information, such as Eq. (7-16), strongly suggests the possibility of wind measurements using the microwave backscattering process.

In the real ocean, the high-frequency waves are not quite noticeable. The most visible feature in the ocean is the large undulation of the water surface with

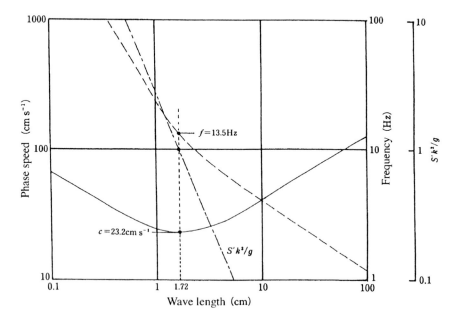

Fig. 7-19. Relationships among the wavelength and the parameters of free surface waves; the phase speed (c), frequency (f) and the restoring force due to the surface tension ($S'k^2/g$).

several tens of meters wavelengths. The energy levels of these water waves are much larger than that of a few centimeter waves. The smaller waves associated with microwave backscattering are riding on the top of those larger waves.

The experimental work of Mitsuyasu and Honda (1974) gave strong support to the principle of microwave wind measurement based on the relationship between the high-frequency water wave energy and the wind. They have measured in a wind-wave facility the high-frequency wind-wave spectrum up to 50 Hz using high-response wave gauges. Their frequency spectra are shown in Fig. 7-20. Focusing attention on the energy level at around 10 Hz band, the α_s of Eq. (7-16) is also a function of u_* and the energy increases with wind. This result provides a basis for the microwave backscattering model (see also Section 2.4.3).

c) Whitecaps and air entrainment

When winds larger than 5 m s^{-1} blow over the ocean, whitecaps become visible (Fig. 7-21). This phenomenon is closely related to the breaking of wind

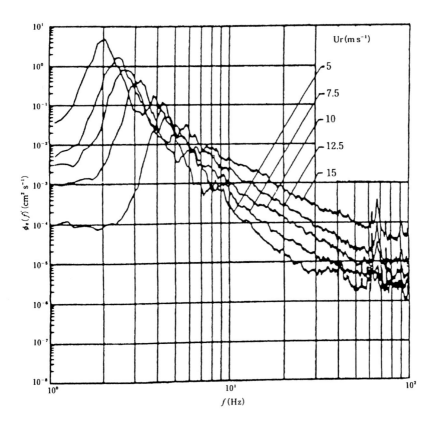

Fig. 7-20. Change of the spectral energy in the high-frequency range due to wind speed (Mitsuyasu and Honda, 1974).

Fig. 7-21. Whitecaps under surface winds of about 10 m/s.

waves, which is a strongly nonlinear phenomenon. When whitecaps are created due to wave breaking, air is entrained into the seawater from the surface.

Since the whitecap coverage percentage is well represented by a function of the breaking wave parameter R_B (Eq. (2-123), Zhao and Toba, 2001) as described earlier in Section 2.5.3, it depends also on the state of wind waves. However, to a first approximation, it increases with the wind. There are a few researches that investigate it quantitatively. Figure 7-22 summarizes the measurements of covering rate. The whitecap coverage increases exponentially with an incease of wind speed. As the whitecap coverage increases, so does the air entrainment.

The sizes of bubbles vary largely in the ocean. Large bubbles rise to the surface due to buoyancy, and when the bubbles break, water droplets are created in the atmosphere. Small bubbles penetrate deeper into the water and stay long, contributing to the air–sea gas exchange. The bubbles near the water surface contribute to microwave radiation from the sea surface.

Processes of bubble entrainment have been studied using their characteristics reflecting acoustic waves in water. Figure 7-23 displays the air bubble distribution near the sea surface measured by sonar placed at the sea bottom. If nothing is included in the seawater, the acoustic waves transmitted from the sea bottom are reflected back at the sea surface. Since the air bubbles resonate with acoustic waves of an appropriate wavelength, they transmit the acoustic waves with the same wavelength. Therefore, if the sonar transmits acoustic waves with a wavelength suitable for bubble detection, one can measure the bubble distribution by detecting the reflected waves. The figure shows a time series of the point measurement of acoustic reflection near the sea

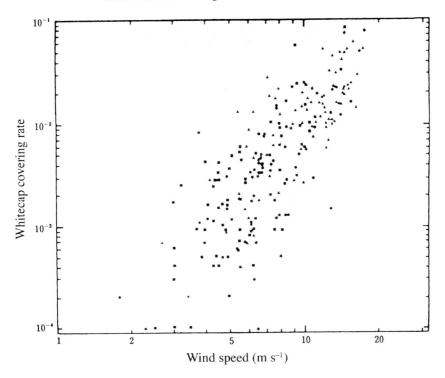

Fig. 7-22. Relationship between aerial coverage of whitecaps and wind speed (Spillane, 1986). The data by Monahan (1971) and Toba and Chaen (1973) are also included.

surface. It can be observed that a large penetration of the bubble clouds occurs sporadically. The air-entrained seawater with bubbles near the sea surface changes the microwave radiation characteristics (Section 7.4.3).

7.4.2 Microwave radiation from the sea surface and the microwave radiometer

a) Microwave radiation from the sea surface without wind

Microwave radiometers measure microwave radiation from the sea surface. Let us consider microwave radiation from the sea surface at thermal equilibrium under the no-wind condition. A radiometer with central frequency f and a narrow bandwidth df receives the radiation expressed as

$$B_b = B_{bf} \cdot df , \qquad (7\text{-}17)$$

where B_b is the brightness temperature. Here B_{bf} (W m^{-2} Sr^{-1}) is spectral brightness temperature at frequency f described by Plank's radiation formula (Eq. (7-1) in frequency). For the microwave range, Plank's formula can be reduced to the Rayleigh–Gene radiation formula because of their low frequencies:

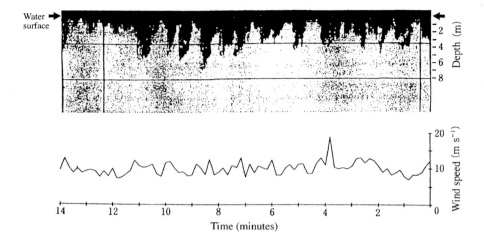

Fig. 7-23. Time series of bubble clouds detected by a sonar placed at the bottom of a lake (the upper panel) and time series of the wind speed observed at the same time (the lower panel) (Thorpe, 1982).

$$B_b = \frac{2kTdf}{\lambda^2}. \qquad (7\text{-}18)$$

The radiation energy per unit time is proportional to the sea surface temperature and inversely proportional to the square of wavelength. The ocean is not a black body for the microwave wavelengths. Using the brightness temperature instead of the equilibrium temperature,

$$B_b = \frac{2kT_B df}{\lambda^2}. \qquad (7\text{-}19)$$

T_B is related to T by the emissivity e of the sea surface:

$$T_B = e(\lambda)T. \qquad (7\text{-}20)$$

When the sea surface temperature is in radiative equilibrium, the microwave emissivity e and the reflection coefficient R satisfy the following relation according to Kirchhoff's radiation law:

$$e_{H,V}(\lambda) = 1 - R_{H,V}(\lambda). \qquad (7\text{-}21)$$

Here, H and V refer to the Horizontal polarization and the Vertical polarization.

The reflection coefficient R can be written as follows from Fresnel's formula, using the complex dielectric constant ε:

$$R_H = \left| \frac{\cos\theta - \sqrt{\varepsilon - \sin^2\theta}}{\cos\theta + \sqrt{\varepsilon - \sin^2\theta}} \right|^2 , \qquad (7\text{-}22)$$

$$R_V = \left| \frac{\varepsilon\cos\theta - \sqrt{\varepsilon - \sin^2\theta}}{\varepsilon\cos\theta + \sqrt{\varepsilon - \sin^2\theta}} \right|^2 . \qquad (7\text{-}23)$$

Here, θ is the incident angle, the angle between the propagation line of electromagnetic waves and the normal to the surface (Swift, 1980). The incident angle dependency of R and e at 10 GHz calculated from Eqs. (7-21) and (7-22) is displayed in Fig. 7-24. For $\theta = 0$, $R_H = R_V = R(0)$, and for 10 GHz ($\lambda = 3$ cm), $R(0) = 0.6$ (see Table 7-4). The emissivity of the sea surface for microwave bands is much smaller than 1 (about 1/3). For calm sea conditions under no wind or weak wind, the microwave radiation from the sea surface is polarized.

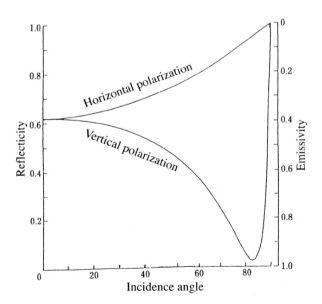

Fig. 7-24. Emissivity and the reflection coefficient of electromagnetic wave of 10 GHz (the wavelength of 3 cm) versus its incidence angle (Stewart, 1985).

Table 7-4. Microwave frequency and polarization of SMMR and SSM/I.

Microwave radiometer	SSMR	SSM/I
Satellite	Seasat	DMSP
Year of launch	1978	1987
Width of swath	600 km	1400 km
Observation frequency/polarization	6.633 V, H	
	10.69 V, H	19.35 V, H
	18.00 V, H	22.235 V
	21.00 V, H	37.0 V, H
	37.00 V, H	85.5 V, H

b) State of the sea surface and microwave radiation from the sea surface

Microwave radiation from the sea surface is closely related to the conditions of sea surface roughened by the surface wind. Microwave radiation from the surface under no wind or weak wind conditions has been described in the previous section. However, in reality, the sea surface is never still even when no wind is blowing. Swell propagates from all directions. As the wind speed increases and the sea surface becomes rough, the microwave radiation increases and the polarization ratio decreases. Figure 7-25 shows observational results from a microwave radiometer installed on a tower using three different wavelengths. It shows the brightness temperatures versus wind speed for vertical polarization, horizontal polarization and their ratios. It is obvious that the horizontal polarization component increases with wind speed. The cause of such behavior is given as follows (Wentz, 1992);

(1) Effect of surface tilt due to surface waves longer than the microwave wavelength.
(2) The effect of bubbles created near the sea surface.
(3) The effect of roughness smaller than the microwave wavelength.

The first effect is caused by the increase of sea surface area radiating microwaves as the wind waves are developed. Stogryn (1967) modeled this effect using a model based on geometrical optics and the statistics of the sea surface slope presented by Cox and Munk (1954a, b). In this model, the sea surface is an ensemble of small tilted elements, and radiation from each element is treated independent of each other. Figure 7-26 shows the wind speed dependency of the horizontal polarization components from 290 K sea surface at 19.4 GHz. For an incident angle larger than 70 degrees, the brightness temperature increases with wind.

When the sea surface becomes rough and the air bubbles are entrained at the surface, the microwave radiation becomes stronger. Smith (1988) has investigated the microwave radiation from nine whitecaps on the sea surface using a radiometer with V- and H-pol 37 GHz and V-pol 19 GHz wavelengths. Figure 7-27(b) is a picture of the whitecap taken from an airplane. The measured microwave signals showed a large increase when the whitecap region was captured.

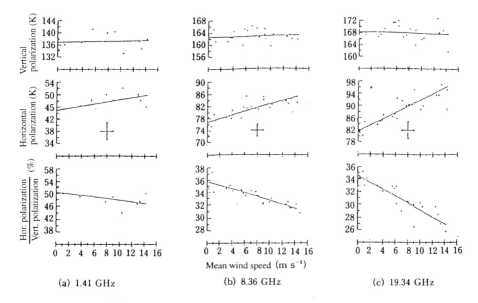

Fig. 7-25. Dependence of the brightness temperatures at the sea surface on the wind speed. The temperatures are measured by a microwave radiometer from the incidence angle of 55°. (a) 1.41 GHz, (b) 8.36 GHz and (c) 19.34 GHz. The upper panels show the component of vertical polarization, the middle the component of horizontal polarization, and (c) the ratio of the vertical- and horizontal-polarization components.

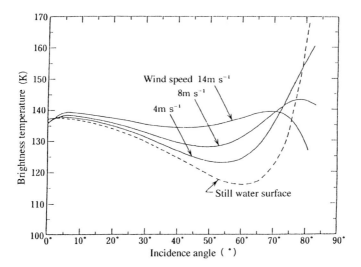

Fig. 7-26. Dependence of the horizontal-polarization component of the brightness temperature on wind speed. The dependence is indicated against the incidence angle (Stogryn, 1967).

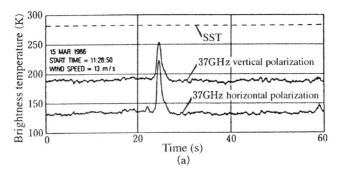

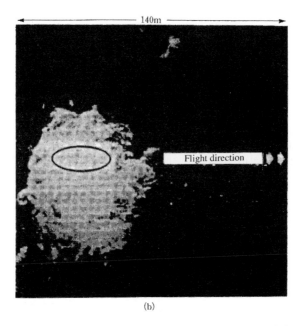

(b)

Fig. 7-27. (a) Change of the 37 GHz brightness temperature observed along the flight path of an airplane, and (b) the picture of a whitecap taken from the airplane. Spikes of the brightness temperature correspond to the ellipse in the whitecap pictured in (b).

Wentz (1975) modeled the microwave radiation from the sea surface, including the effects of surface configurations with wavelengths longer and shorter than the microwave wavelength and the effect of scattering of the sky radiation at the surface. The roughness elements shorter than the microwave wavelength are tilted by the longer wind waves. The latter effect explained better the result by Hollinger (1971) than just taking into effect (1), but their contributions were small compared to the other effects.

c) Simultaneous measurement of microwave radiation with multiple frequency and multiple polarization

The brightness temperatures measured by a microwave radiometer contain rich information about the physical quantities for both the atmosphere and the ocean. In order to retrieve a physical parameter, for example wind speed, single wave or single polarization measurement is not sufficient. To separate out different factors from the radiometer measurements through dependence of physical parameters on the microwave radiation in different frequency and polarization, simultaneous observation with multiple frequency and polarization is carried out. In other words, the measurement is conducted with the combination of the most and the least sensitive wavelength and polarization for the targeting parameter.

Figure 7-28 shows the changing rate of brightness temperatures measured with different frequencies against the physical variables (P_I) (Wilheit *et al.*, 1980). The arrow indicates the frequencies of the Scanning Multichannel Microwave Radiometer (SMMR) on board Seasat. The wind wave dependency increases with frequency up to 10 GHz, and remains constant beyond that. The frequencies of 6.6 GHz and 10.7 GHz are main contributors for retrieval of SST and surface wind, respectively, and 21 GHz and 37 GHz are used for vapor and cloud corrections, respectively. The frequency of 18 GHz does not have its own observation purpose, but is added in order to improve the accuracy by stabilizing the parameter retrieval scheme.

d) Microwave radiometer

Microwave radiometers usually conduct earth observation using multiple

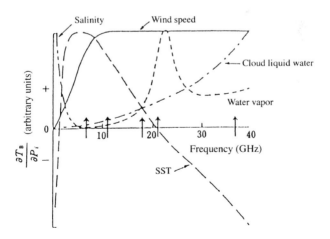

Fig. 7-28. Dependence of changing rate of the brightness temperatures on the physical parameters. The arrows indicate the frequencies observed by the SMMR (Wilheit *et al.*, 1980).

frequencies and polarities. Test sensors were on board Nimbus 5 and 6 and Skylab. The first operational oceanographic sensor was the SSMR on board Seasat (Njoku *et al.*, 1980). The SMMR had 5 frequencies, and 2 polarizations, a total of 10 channels, measuring sea surface temperature, wind speed, vapor content and cloud water content in the marine atmosphere. It was then succeeded by the SSM/I (Special Sensor Microwave Imager, microwave radiometer) on board the DMSP (Defense Meteorological Satellite Program). Both the SMMR and the SSM/I frequencies and polarization are shown in Table 7-4. The SSMR had a fixed grazing angle at 50 degrees optimum for ocean observation, and observed microwave radiation by the conical-scan method.

The SMMR had a few problems: a cross talk of different signals and degradation of the observational signals due to complicated calibration and retrieval methods. The SSM/I on board the DMSP resolved these problems (Hollinger *et al.*, 1987). There are seven channels for four wavelengths and different polarization. Because it does not have a 6.6 GHz band, SST cannot be retrieved. The SSM/I has an 85.5 GHz channel that measures snow and ice with high spatial resolution.

7.4.3 Microwave backscattering from sea surface, and scatterometer and radiometer

a) Microwave backscattering coefficient σ_0 at the sea surface and the ocean wind

Satellite-borne microwave scatterometers and altimeters transmit microwave pulses to the ocean and obtain information on the wind speed from the signals backscattered at the sea surface. Therefore, one needs to know the relation between the microwave backscattering coefficient σ_0 and the wind *a priori*. One way to understand the characteristics of microwave backscattering at the sea surface is to solve Maxwell's equations with the boundary conditions of sea surface geometry. However, this is almost impossible since the sea surface changes its shape with time in a rather complex manner. It was consequently considered to describe the sea surface configuration with a statistical quantity and to model the backscattering coefficient as a function of that quantity. In the case in which the wind speed is the target parameter, the relation between the wind speed and the sea surface configuration must be known *a priori*.

When the wind blows in a wind-wave tunnel, the water surface undulates and the surface is covered with small roughness. The microwave backscattering coefficient can be obtained as a function of wind by illuminating such surface with a microwave radar and measuring the reflected electromagnetic waves (Fig. 7-29). The received electromagnetic power (equivalent to σ_0) changes with wind speed and the incident angle θ, which is measured from the normal incidence (Fig. 7-30(a)). $\theta > 0$ means that the radar antenna points upwind. When the incidence angle is 0 degrees, σ_0 decreases with wind. For $|\theta| > 20$ degrees, σ_0 increases with roughly the square of the wind speed. From either $\sigma_0 - u$ relation, the wind speed can be estimated. The case of 0-degree incidence angle corresponds to the altimeter measurement. The scatterometer principle of wind measurement is based on the relationship for incidence angles larger than 20 degree.

b) Specular reflection model
    Let us consider microwave reflection from a flat sea surface covered with small-scale roughness, the wavelengths of which are smaller than the microwave

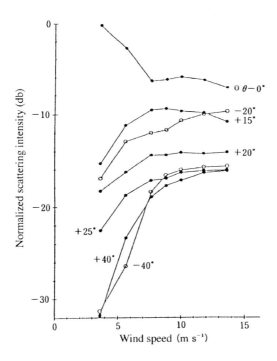

Fig. 7-29. Dependence of normalized backscattering intensity of the 10 GHz microwave on the wind speed and incidence angle (Ebuchi *et al.*, 1992).

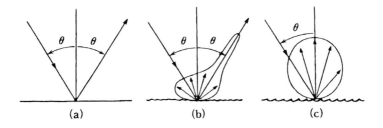

Fig. 7-30. Reflection of the microwave at the surface covered by the roughness with wavelengths shorter than that of the microwave. (a) Smooth surface, (b) slightly rough surface, and (c) rough surface.

wavelength. Figure 7-30 is a schematic describing the reflection pattern associated with changes in the roughness. Figure 7-30(a) shows the reflection pattern for a flat surface, which is called specular reflection and can be described by Fresnel's formula (Fig. 7-24). Figure 7-30(b) is for a surface with intermediate roughness. In this case, the specular reflection reduces compared with (a), and there are components that get scattered in all directions. These are called diffused component or incoherent component. The component of specular reflection is coherent. When the roughness increases, the reflection becomes purely diffusive (Fig. 7-24(c)). Such a surface is called Lambertian. The sea surface covered with wind waves is considered to be in the state described in Fig. 7-30(b).

Now, we extend the microwave scattering at the roughened horizontal surface to the roughened wavy surface. The most fundamental method to relate the sea surface configuration and σ_0 is to use the geometrical optics approximation as shown earlier in Fig. 7-30(a). Assuming that the radius of curvature of the sea surface is much larger than the microwave wavelength, the sea surface can be modeled as an ensemble of small facets, which are mirror-like microwave reflectors. This model is called a specular point model.

Brrick (1968) has described σ_0 by the probability density function of the surface slope $p(\eta_x, \eta_y)$

$$\sigma_0 = \pi \sec^4 \theta \cdot p(\eta_x, \eta_y) \cdot |R(0)|^2 . \tag{7-24}$$

Here, θ is the incident angle, η_x, η_y are the slopes in the x- and y-directions, $R(0)$ is the Fresnel reflection coefficient at nadir which can be expressed by Eqs. (7-21) and (7-22). When the sea surface slope is not directional and Gaussian (Cox and Munk, 1954a; see Section 7.4.2), Eq. (7-24) becomes

$$\sigma_0 = \sec^4 \theta \cdot \left(\frac{|R(0)|^2}{s^2} \right) \exp\left(-\frac{\tan^2 \theta}{s^2} \right). \tag{7-25}$$

Here $s^2 = \eta_x^2 + \eta_y^2$. Cox and Munk found the following relation between s^2 and the wind speed at 10 m, U_{10}:

$$s^2 = 0.003 + 0.00512 U_{10} . \tag{7-26}$$

This formula agrees quite well with the observed σ_0 (Valenzuela, 1978).

c) Resonant scattering model

As shown earlier in Fig. 7-29, σ_0 increases with wind speed, as the incident angle of the microwave increases. In the ocean, it is known that the sea surface slopes hardly exceed 20 degrees. Therefore, when the incidence angle is larger than 20 degrees, the specular point model would not work. In other words, the microwaves incident at $|\theta| > 20$ will be reflected in the direction opposite to the antenna according to the specular point model (Fig. 7-30(a)). This is not in agreement with the experiment (Section 7.4.4(a)).

When the sea surface is rough, the scattered components propagating along the antenna direction are much smaller than the specular reflected components. Wright (1966) has studied microwave backscattering from water waves using artificially generated wind waves in a water tank, and found the following relation among the following quantities: water-wave wavelength propagating at the direction of the microwave incidence, the microwave wavelength, and the incident angle (Fig. 7-31):

$$k_w = 2k_e \cdot \sin\theta . \qquad (7\text{-}27)$$

Here, k_w is the water wave wave-number ($k_w = 2\pi/\lambda_w$, λ_w is the water wave wavelength), k_e is the microwave wavelength. The V-pol backscattering is larger than that of the H-pol, and more so for larger incidence angle. The observed backscattering is explained by the first order perturbation scattering theory. There is a resonance between the incident microwave and the scattered microwave in the antenna direction. According to scattering theory, the cross section of the backscatter can be expressed as (Valenzuela, 1978)

$$\sigma_0(\theta)_{ij} = 4\pi k_e^{\,4} \cdot \cos^4\theta \cdot \left| g_{ij}(\theta) \right|^2 \cdot W(2k_e \cdot \sin\theta, 0) . \qquad (7\text{-}28)$$

Here $W(k, \phi)$ is the two-dimensional wave number spectral density of the sea surface elevation. In this formula, g_{ij} is the first order scattering coefficient and the suffixes, i and j, indicate the polarization of the incident and backscattered wave respectively:

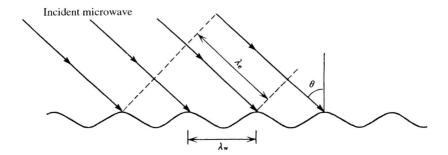

Fig. 7-31. A schematic of resonant scattering.

$$g_{HH}(\theta) = \frac{\varepsilon_r - 1}{\left[\cos\theta + \left(\varepsilon_r - \sin^2\theta\right)^{1/2}\right]^2},$$ (7-29)

$$g_{VV}(\theta) = \frac{\left(\varepsilon_r - 1\right)\left\{\varepsilon_r\left(1 + \sin^2\theta\right) - \sin^2\theta\right\}}{\left[\varepsilon_r\cos\theta + \left(\varepsilon_r - \sin^2\theta\right)^{1/2}\right]^2}.$$ (7-30)

Here, ε_r is the dielectric constant of seawater.

The condition (7-26) resembles the diffraction relation between the X-ray wavelength and the geometry of repeating lattice found in the early 20th century. Since the wavelengths of microwave and water wave are related through Bragg's equation, this mechanism is called Bragg scattering or resonance scattering.

In this mechanism, the σ_0 is related to the wave-number spectrum. For example, 10 GHz microwave has a wavelength of about 3 cm, so the water waves that can resonate with this microwave have the wavelengths of about a few centimeters and are influenced by not only gravity but also surface tension (Fig. 7-19). There are some unknowns in the spectra in this wave number range, which are mostly due to the difficulty in measuring the spatial distribution of surface elevation with a few centimeter scales. In contrast, it is easier to obtain the frequency spectra, from which the wave-number spectra $W(k, \phi)$ can be derived assuming a linear dispersion relation and spatial homogeneity. In the case of the Phillips spectrum with a frequency range in which the spectrum energy is proportional to the −5th power of frequency:

$$W(k, \phi) = B_p k^{-4};$$ (7-31)

In the case of the Toba spectrum with a frequency range in which the spectrum energy is proportional to the −4th power of frequency:

$$W(k, \phi) = B_T u_* k^{-3.5}.$$ (7-32)

In these formulae, B_p and B_T are constants. In the Phillips spectra, the spectrum energy of surface waves that interact with the microwave do not depend on the wind. On the other hand, in the Toba spectra, the spectral energy in the high frequency range is proportional to the friction velocity.

From a precise measurement of surface elevation, Mitsuyasu and Honda (1974) have shown that, for high frequency range between 10–50 Hz, B_p is proportional to u_* to the power of 9/4 (=2.25) and B_T to u_* to the power of 1.5. For the Toba spectra, adding the 1.5th power of u_* to u_* explicitly included in the formula, the spectral level is nearly proportional to the 2.5th power of u_* (Eq. (7-32)).

As mentioned earlier, σ_0 is nearly proportional to the square of wind speed. The finding by Mitsuyasu and Honda (1974) that the $W(k, \phi)$ level depends on nearly the square of wind speed became evidence of the theoretically derived formula (7-27) for the σ_0 dependence on wind. It is a great contribution to the satellite-borne radar measurement of wind speed.

Jones and Schroeder (1978) have summarized the wind speed dependence of σ_0 as

$$\sigma_0 = aU^x \quad \text{or} \quad \log \sigma_0 = a' + x \log U . \tag{7-33}$$

U is the wind speed at a certain height and a and a' are constants. The validity of this formula has been verified by various experiments. The exponent x varies with the frequency of the electromagnetic wave and ranges from 0.5 to 2.0 (Fig. 7-32, Jones and Schroeder, 1978; Stewart, 1985). The value of x is larger for shorter electromagnetic wavelengths. Therefore, short waves are more sensitive and suitable for wind speed measurement over the ocean.

d) Composite scattering model

In the ocean surface, waves from a few centimeters to a few decimeters, which contribute to resonance scattering, exist together with the main longer waves, creating spectrum energy distribution in a wide range of frequencies. The main energy contributor can be either wind waves or swells, depending on the

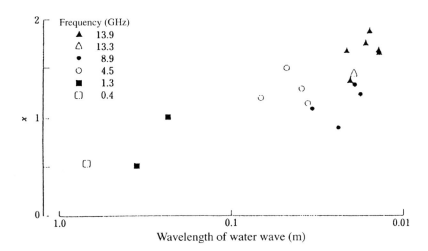

Fig. 7-32. The radar scattering cross section varies as wind speed to some power χ, which is shown as a function of wavelength of the surface waves scattering the microwaves.

weather situation. In order to model the backscattering from a realistic ocean surface, Wright (1968) considered a case in which the short waves responsible for the resonant scattering are tilted by the underlying long waves. This is called a composite model or two-scale model.

Figure 7-33 shows the comparison of the experimental data taken in the oceans and the microwave backscattering models (Valenzuela, 1978). For small incidence angle, the specular point model is used (broken line), and for larger incidence angle, the composite scattering model explains well the observations. Now, the specular point model for a small incident angle and the composite model for a large incidence angle have been well accepted.

e) Directionality of the microwave backscattering

The σ_0 for a microwave with the wavelengths of a few centimeters changes with the azimuth angle ϕ between the wind direction and the microwave incident direction. The σ_0 becomes the largest when the microwave incident direction is upwind ($\phi = 0$) and downwind ($\phi = 180$), and becomes the smallest when the microwave direction is normal to the wind ($\phi = 90, 270$). In general, the maximum in the upwind direction is larger than the maximum in the downwind direction, but the difference becomes small as the wind speed increases. The σ_0 with directionality can be expressed as

$$\sigma_0 = aU^x(1 + b\cos\theta + c\cos 2\theta). \qquad (7\text{-}34)$$

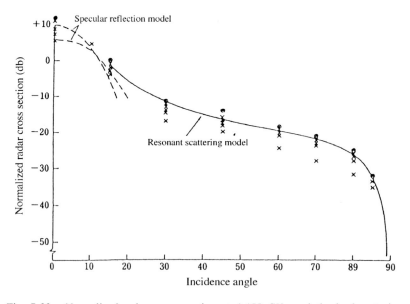

Fig. 7-33.   Normalized radar cross section at 4.455 GHz and the backscattering mechanisms at various incidence angles (Valenzuela, 1978).

Here, b, c, are empirical constants. An example of the directionality of σ_0 is presented in Fig. 7-34 (Jones and Schroeder, 1978).

If the σ_0 at a single location can be measured simultaneously from different angles, the wind direction can be determined using the relationship shown in Fig. 7-34. A schematic describing the algorithm for determination of wind speed and wind direction determination through Eq. (7-34) is shown in Fig. 7-35 (Jones et al., 1982). Suppose an observation is made to obtain $\sigma(\theta)$. From Eq. (7-33), the combination of the wind speed and direction to realize $\sigma(\theta)$ is infinite. However, once the observation at $\theta + 90$ degrees, $\sigma(\theta + 90)$, is given, the possible combination of the wind speed and direction reduces to four cross points of $\sigma(\theta)$ and $\sigma(\theta + 90)$ as shown in Fig. 7-34. The SASS on board the Seasat observed the ocean surface from these two angles and the wind speed and the wind direction was determined using additional information such as the atmospheric pressure distribution.

f) Microwave scatterometer

The SASS on board the Seasat is a microwave scatterometer operated at 14.6 GHz (Ku-band). It had two antennas set at an angle of 90 degrees on each side for measurement of the surface wind vectors in the two swaths of about 600 km along the satellite pass. The area of 400 km right under the satellite cannot be measured. This is because the measurement principle of the scatterometer is based on backscattering at a grazing incident.

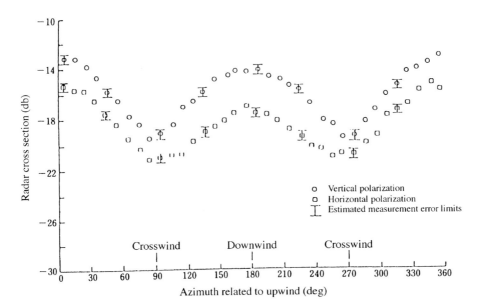

Fig. 7-34. Dependence of the normalized radar cross section on the angle relative to the mean wind for 13.9 GHz at 40 degree incidence angle (Jones and Schroeder, 1978).

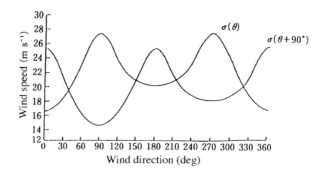

Fig. 7-35. A schematic diagram on determination of wind speed and wind direction using the normalized cross sections measured from different azimuthal angles (Jones *et al.*, 1982).

In the case using two antennas to observe a point from two directions, there are four ambiguous solution (Section 7.4.4(e)). Several algorithms have been proposed to determine the proper wind direction from the SASS observations. However, they necessitate the use of meteorological data or numerical models besides the SASS data and their applicability was limited. The AMI (Active Microwave Instrument) aboard ERS-1 (European Remote-sensing Satellite 1) and the NSCAT (NASA Scatterometer) aboard the ADEOS (Advanced Earth Observing Satellite) uses three antennas looking at the same point in order to improve the measurement accuracy and remove the ambiguity in determination of the wind direction.

g) Microwave altimeter

A microwave altimeter measures the undulation of the sea surface with a scale of 100 km in the horizontal direction. The propagation time of a microwave pulse transmitted directly under the satellite and reflecting back from the sea surface is measured and converted to the range between the satellite and the surface through many corrections. The sea surface heights can be used to obtain geostrophic currents in the surface layer of ocean. The main target of this instrument is to measure the surface currents in geostrophic balance in the world oceans.

In addition, the strength of the received signal and the deformation of the pulse shape contain valuable information about the sea surface. The measurement principle is described in Sections 7.4.4(a) and (b) for nadir ($\theta = 0$) angle, and the wind speed can be measured from σ_0. The σ_0 decreases with the wind speed, and the wind direction cannot be measured.

The deformation of the pulse shape is caused by the configuration of the dominant surface wave, which enables us to measure the significant wave height.

7.5 AN EXAMPLE OF THE AIR–SEA INTERACTION STUDY BY SATELLITE OBSERVATION

Satellite remote sensing provides a valuable tool for gathering huge amount of information about the ocean surface and the atmosphere over it. It could be a powerful tool for air–sea interaction research but the methodology is yet to be established. Below, several examples of air–sea interaction research using satellite data are introduced.

7.5.1 The growth of wind waves in the Sea of Japan during winter monsoon

As mentioned earlier in Section 7.1.2(b), when the winter monsoon wind starts blowing, there is an active air–sea interaction in the Sea of Japan. The cold wind from the continent influences the entire Far East Asia region and is a significant factor that governs the climate of Japan. However, because observations need to be made in the sea during severe conditions, investigation of the air-sea interaction is not fully conducted yet. Here we introduce a satellite observational study investigating such a case (Ebuchi et al., 1992).

Figure 7-36(a) is a weather chart of 10 January 1987 when the monsoon wind started to blow. The trajectory of the Geosat (Geodetic Satellite) over the Sea of Japan during the same period is shown as well. GEOSAT had a microwave altimeter and measured both the wind and the waves beneath it (Section 7.4.4(g)). The profiles of the significant wave height and wind speed are drawn in the figures right of the Geosat trajectory in Fig. 7-36(b). The chart of atmospheric pressure and the wind speed from the altimeter (Fig. 7-36(b) center) shows that the wind of 10–20 m s^{-1} blows from the Asian continent to Japan. Under strong winds, wind waves develop due to increase of the fetch, which is the distance from the coastline along the wind streamline (Fig. 7-36(b) left). The wave height is about 5 m around the center in the Sea of Japan. In general, it is difficult to conduct ship observations during such a strong winter wind condition, though there are many research subjects to pursue since the strong air–sea interactions involves, exchanging of materials, momentum and heat energy between the air and the sea.

The development of wind waves with fetch can be expressed by a so-called fetch graph (Wilson, 1965). The fetch and the wave height are non-dimensionalized by the wind speed and plotted in the logarithmic diagram. Figure 7-37 is a fetch graph drawn from the wave height and wind speed obtained by the GEOSAT altimeter. Sixteen cases similar to that shown in Fig. 7-36 were collected and analyzed; the fetch is determined using the streamline given by the weather chart. The plots agree well with the fetch laws presented earlier by JONSWAP (Joint North Sea Wave Project; Hasselmann et al., 1973) and Wilson (1965). In order to generate such figures, a well-organized and extensive observational project like JONSWAP was necessary in the past.

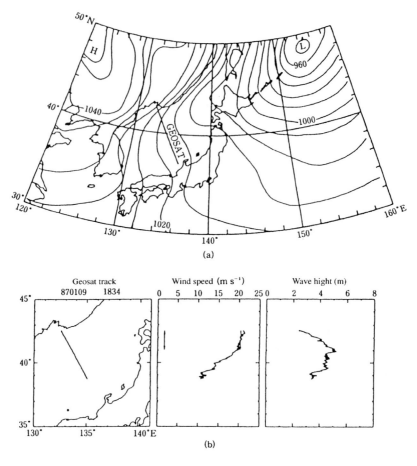

Fig. 7-36. (a) Surface pressure on 9 January 1987 and the grand track of the GEOSAT, and (b) profiles of the wind speed and the wave height observed by the GEOSAT along the track shown in (a) (Ebuchi *et al.*, 1992).

7.5.2 *Estimation of latent heat flux by satellite observations*

Turbulent heat fluxes at the sea surface are a driving force for the motions of the atmosphere and the ocean, as in the case of momentum flux. The turbulent heat flux in most of the open oceans is contributed by the latent heat flux described by Eq. (7-5). Liu (1984) used the SMMR data from the Seasat, and estimated the latent heat flux in the open ocean. Relative humidity in the air near the sea surface was estimated from the SMMR-derived total vapor content using the statistical relation between relative humidity q_a and the total vapor contents of the atmosphere. From the SMMR-derived SST, the saturated relative humidity at the sea surface is obtained. Combining the humidity in the air and at the surface and the wind speed from the SMMR, the latent heat flux can be estimated

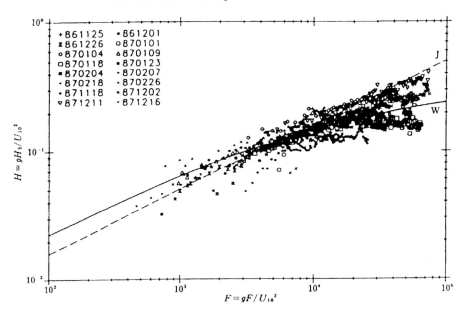

Fig. 7-37. Fetch graph drawn using wind speed and wave height observed by the GEOSAT. Empirical formula proposed by Wilson (W in the figure) and JONSWAP (J) are indicated for comparison (Ebuchi *et al.*, 1992).

through Eq. (7-5). A comparison of the latent heat fluxes derived from ship observations and the Seasat in the North Atlantic Ocean was made for three months (corresponding to the operational period of the Seasat) for 2-degrees grid, and the standard variation was 34.9 W m^{-2}.

Sea surface fluxes estimated in the sea south of Japan are shown below (Kawamura *et al.*, 1993c). Figure 7-38 is the SST distribution in the area estimated using the MCSST method (Section 7.3.4) from the NOAA AVHRR data. The wind velocity at 10 m height is obtained from the microwave radiometer SSMI (Section 7.4.3(d)). Sea surface humidity q_a is estimated through an empirical formula relating the total vapor content from SSMI, and the *in situ* relative humidity q_s. Figure 7-38 shows the daily average of latent heat flux for the 0.25 degrees grid within the 10 degrees square south of Japan (Fig. 7-39). One may observe the following in the SST graph. The Kuroshio current is characterized by a dark band with the highest temperature. In a lower temperature region south of the Kuroshio, the mixed layer has developed during winter. The subtropical front including many eddies are seen at around 25 degrees latitude. The Kuroshio has a large meander near Kii-peninsula and a warm water mass is observed to extend from the crest of the meander at around (31°N, 138°E). This is caused by an outbreak of the Kuroshio warm core (Toba *et al.*, 1991). From Fig. 7-39, one can observe a large latent heat flux corresponding to the Kuroshio and the outbreak region. There is a region where the agreement is poor, which is mainly

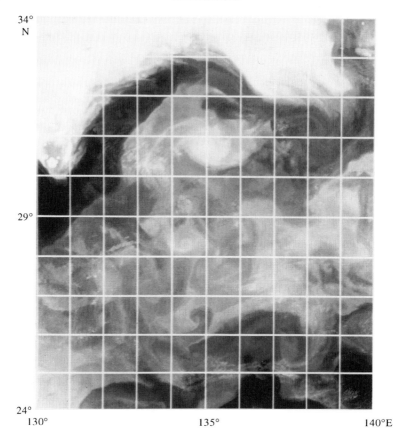

Fig. 7-38. MCSST distribution in the ocean south of Japan derived from AVHRR infrared measurement on 13 May 1998 (Kawamura *et al.*, 1993c).

due to the wind distribution. Wind speed is the main factor to control the latent heat flux and SST is second because it is included in Eq. (7-3) through the relative humidity. Figure 7-39 captures the heat flux distribution influenced by both the wind and SST fields.

7.5.3 Solar radiation estimate

Solar radiation can be monitored once an hour using visible observation by the geostationary meteorological satellite (Section 7.2.4). Figure 7-40 displays the daily average of the solar radiation distribution in the sea south of Japan derived from the VISSR data aboard GMS 3. The sky was mostly clear on this particular day, and thus large values are observed in the whole region. There are a few spots of the smaller solar radiation due to clouds.

In this region, the daily-averaged solar radiation obtained by the satellite is

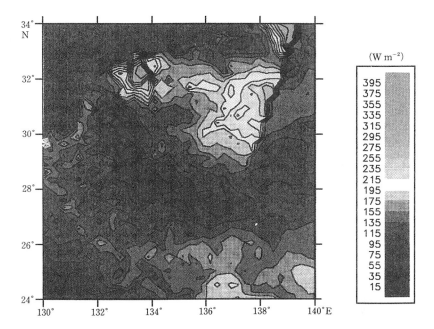

Fig. 7-39. Distribution of latent heat flux in the ocean south of Japan on 13 May 1988. High flux region is seen in the northern area corresponding to the Kuroshio axis shown as the dark band in Fig. 7-38, and in the northeast to the central area corresponding to the strong winds.

compared with those observed by the research vessels for April and May 1988 (Fig. 7-41). The RMSD (Eq. (7-13)) is 14 W m^{-2} (Kawamura *et al.*, 1993c).

7.5.4 Ocean wind measured by the microwave scatterometer during "Yamase"

"Yamase" is the name of a local easterly wind blowing from the Pacific Ocean off the Sanriku Coast to the Tohoku district during summer (Kawamura, 1995b). The wind is cold and humid, accompanying a low cloud. The Yamase phenomena in the summer of 1993 caused extremely cold weather in northern Japan and damaged the rice crops in this region. The Yamase clouds observed by the AVHRR (Section 7.3.2) is shown in Fig. 7-42. The white indicates the clouds and the black the ocean. The gray area shows the lands of Japan and Russia. The mountains higher than 100 meters in the Kitakami heights block the Yamase clouds.

ERS-1, which was an earth observation satellite operated by the European Space Agency, was launched in 1991 carrying a microwave scatterometer AMI (Section 7.4.4(f)). The ocean wind vector observed by the AMI is superimposed in Fig. 7-42. The ocean wind vectors, in general, indicate a northeast wind, but near the Kitakami heights, splits into north and south directions. The northern

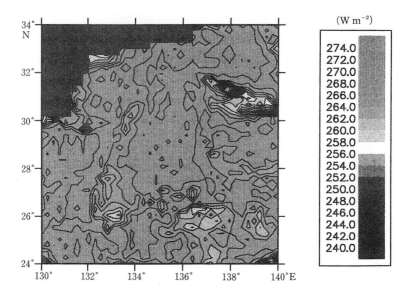

Fig. 7-40. Distribution of the daily solar radiation at the sea surface in the ocean south of Japan (Kawamura *et al.*, 1993c).

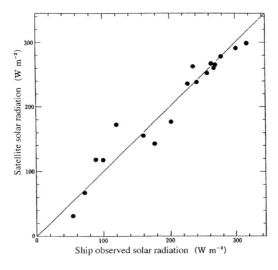

Fig. 7-41. Comparison of the daily solar radiation derived from the VISSR and observed by research vessels in the ocean south of Japan during April and May 1998.

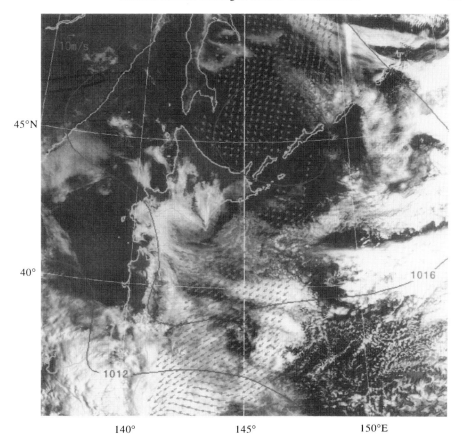

Fig. 7-42. AVHRR visible image capturing the Yamase cloud on 6 August 1993. The ocean surface winds by the AMI on board ERS-1 are superimposed (Kawamura, 1995c).

route carries the Yamase clouds to Aomori and Southern Hokkaido. This illustrates the important role of the Kitakami heights, which induce topographical forcing on the Yamase wind in the northern Tohoku region (Kawamura, 1995c).

The generation and development process of the Yamase phenomena takes place in the vast oceans extending northeast of northern Japan. The *in situ* ocean observations are limited and thus the observational points are much fewer than those on the ground. The measurement of the wind field in the ocean, which was difficult in the past, is now improving available and providing new understanding.

7.5.5 MCSST and the sudden change of the SST during typhoon passage

Through an appropriate atmospheric correction, SST can be measured with high accuracy from the infrared observation of the AVHRR (Section 7.3.4). The sudden change in the SST observed by such measurement is indicated below.

Also, the accuracy of the satellite observed physical quantities is discussed.

Two typhoons passed in the Pacific Ocean off northern Japan from 15–18 August 1989. The typhoon paths are shown in Fig. 7-43. In the same figure are shown the front of the Kuroshio extension and the location of the warm core ring. Figure 7-44 shows the MCSST distribution before the passage of the typhoon (a) and after (b). Their difference is illustrated in Fig. 7-44(c). The images before and after the typhoon passage were created from several AVHRR data, removing the clouds. Due to the passage of the typhoons, there was a large SST cooling in the area north of the Kuroshio front and 36°N. Furthermore, the warm core ring and the Kuroshio front became more visible. Figure 7-43 is drawn based on the SST image taken after the passage of the typhoons (Fig. 7-44(b)).

The RMSD is often used for evaluating the quality of satellite-derived physical parameters (Section 7.3.4(b)). The RMSD of MCSST is about 0.6°C. When the spatial variation or the temporal variation of SST is larger than 0.6 K, the variation is statistically reliable. The temperature change caused by a typhoon passage in the central to the northern part of Fig. 7-44(c) is a strong signal far

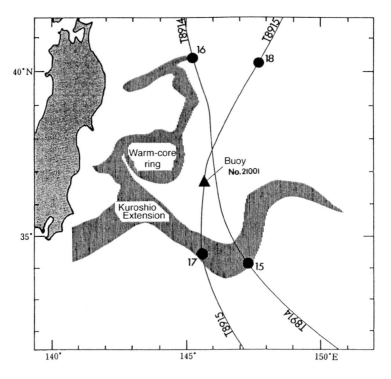

Fig. 7-43. Tracks of the two typhoons passing over the Pacific Ocean off the Tohoku region of Japan in August 1989. Oceanic structures of the warm-core ring and the Kuroshio Extension are schematically indicated. The typhoons passed during 15–18 August (Sakaida et al., 1998).

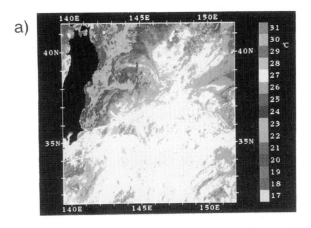

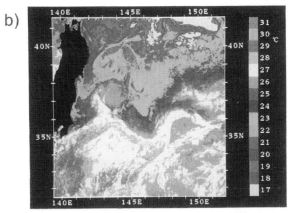

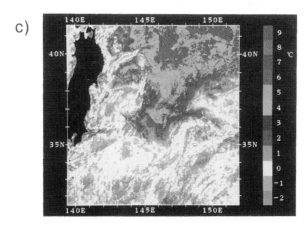

Fig. 7-44. MCSST distributions (a) before and (b) after the typhoons passage. (c) Difference of the MCSST distributions before and after the typhoons passage. Positive difference means SST cooling (Sakaida *et al.*, 1998).

exceeding that.

The oceanic region with large SST cooling is limited to the Oyashio region and spreads widely in space. The Oyashio water is covered by a thin seasonal mixed layer during summer. When the typhoon passed by, the strong mechanical mixing destroyed the thin mixed layer and the cold water outcropped (Sakaida *et al.*, 1998).

REFERENCES

Anding, D. and R. Kauth (1970): *Remote Sens. Environ.*, **1**, 217–220.

Barrick, D. E. (1968): *IEEE Trans. Antennas and Propagation*, **AP-16**, 449–454.

Bishop, J. K. and W. B. Rossow (1991): *J. Geophys. Res.*, **96**(C9), 16839–16858.

Cox, C. and W. Munk (1954a): *J. Opt. Soc. Amer.*, **44**, 838–850.

Cox, C. and W. Munk (1954b): *J. Marine Res.*, 198–227.

Ebuchi, N., H. Kawamura and Y. Toba (1992): *J. Geophys. Res.*, **92**, 809–819.

Freilich, M. (1985): Science opportunities using the NASA scatterometer on N-ROSS. JPL Publication 84-57, 36 pp.

Gautier, C., G. Diak and S. Masse (1980): *J. Appl. Met.*, **19**, 1005–1012.

Hasselmann, K. *et al.* (1973):*Dtsch. Hydrogr. Z.*, *8*(12), Suppl. A., 95 pp.

Hollinger, J. P. (1971): *IEEE Trans. Geoscience Elect.*, **GE-9**, 165–169.

Hollinger, J., R. Lo, G. Poe, R. Savage and J. Peirce (1987): Special sensor microwave/imager user's guide. Naval Res. Lab., 119 pp.

JMA (eds.) (1984): *Application of Meteorological Satellite Data in Weather Forecasting*, Japan Weather Association, 322 pp. (in Japanese).

Jones, W. L. and L. C. Schroeder (1978): *Boudary-Layer Meteorol.*, **13**, 133–149.

Jones, W. L., L. C. Schroeder, D. H. Boogs, E. M. Bracalente, R. A. Brown, G. J. Dome, W. J. Pierson and F. J. Wentz (1982): *J. Geophys. Res.*, **87**(C5), 3297–3317.

Katsaros, K. B. (1980): *Boubndary-Layer Meteorol.*, **18**, 107–127.

Kawamura, H. (1995a): *Tenki*, **42**(6), 347–354 (in Japanese).

Kawamura, H. (ed.) (1995b): *Kishou-Kenkyu Note*, **183**, 179 pp. (in Japanese).

Kawamura, H. (1995c): *Kishou-Kenkyu Note*, **183**, 153–179 (in Japanese).

Kawamura, H., S. Kizu, F. Sakaida and Y. Toba (1993a): *Tohoku Geophysical Journal* (The Science Reports of the Tohoku University, Series 5), **36**, 89–102.

Kawamura, H., F. Sakaida and S. Kizu (1993b): *Tohoku Geophysical Journal* (The Science Reports of the Tohoku University, Series 5), **36**, 103–114.

Kawamura, H., S. Kizu, F. Sakaida, N. Ebuchi and Y. Toba (1993c): *Remote Sensing of the Oceans*, ed. by I. S. F. Jones *et al.*, pp. 263–270.

Kimura, K. and N. Murayama (1990): *Kishou-Kenkyu Note*, **169**, 5–22 (in Japanese).

Kinsman, B. (1965): *Wind Waves—Their Generation and Propagation on the Ocean Surface*. Prentice-Hall Inc., Englewood Cliffs, N.J., 676 pp.

Kizu, S., H. Kawamura and Y. Toba (1990): *Gogai-Kaiyo*, 17–21 (in Japanese).

Kondo, J. (ed.) (1994): *Meteorology of Water Environment*, Asakura-Shoten, 348 pp. (in Japanese).

Liu, W. T. (1984): *Large-Scale Oceanographic Experiments and Satellite*, ed. by C. Gautier and M. Fieux, D. Reidel Publishing, pp. 205–221.

Mitsuyasu, H. and T. Honda (1974): *J. Oceanogr. Soc. Japan*, **30**, 185–198.

Monahan, E. C. (1971): *J. Phys. Oceanogr.*, **1**, 139–144.

Njoku, E. G., J. M. Stacey and F. T. Barath (1980): *IEEE J. Oceanic Eng.*, **OE-5**, 100–115.

NOAA (1991): *NOAA Polar Orbiter Data Users Guide*, Kidwell, K. B., NOAA/NESDIS/NCDC/SDSD, 187 pp.

Phillips, O. M. (1958): *J. Fluid Mech.*, **4**, 426–434.

Rashke, E., A. Gratzki and M. Rieland (1987): *J. Climatol.*, **7**, 205–213.

Rossow, W. B. and R. A. Schiffer (1991): *Bull. Am. Meteorol. Soc.*, **72**, 2–20.

Sakaida, F. and H. Kawamura (1992): *J. Oceanogr.*, **48**, 179–192.

Sakaida, F., H. Kawamura and Y. Toba (1998): Sea surface cooling caused by typhoons in the Tohoku Area in August 1989, *J. Geophys. Res.*, **103**, 1053–1065.

Sasaki, H., N. Murayama and K. Arai (1990): *Kishou-Kenkyu Note*, **169**, 61–88 (in Japanese).

Saunders, P. M. (1967): *J. Atmos. Sci.*, **24**, 269–273.

Smith, P. M. (1988): *IEEE Trans. Geoscience and Remote Sensing*, **GE26**, 541–547.

Spillane, E. C., E. C. Monahan, P. A. Bowyer, D. M. Doyle and P. J. Stabeno (1986): *Oceanic Whitecaps*. ed. by E. C. Monahan and G. MacNiocaill, pp. 209–218.

Stewart, R. H. (1985): *Methods of Satellite Oceanography*. University of California Press, 360 pp.

Stogryn, A. (1967): *IEEE Trans. Antennas and Propag.*, **AP-5**, 278–286.

Swift, C. T. (1980): *Boundary-Layer Meteorol.*, **18**, 24–54.

Thorpe, S. A. (1982): *Phil. Trans. Roy. Soc. London*, **A-304**, 155–210.

Toba, Y. (1973): *J. Oceanogr. Soc. Japan*, **29**, 209–220.

Toba, Y. and M. Chaen (1973): *Rec. Oceanogr. Works Japan*, **12**, 1–11.

Toba, Y., H. Kawamura. K. Hanawa, H. Otobe and K. Taira (1991): *J. Oceanogr. Soc. Japan*, **47**, 297–303.

Uda, M. (1978): *History of Ocean Research Evolution*, Tokai University Press, 331 pp. (in Japanese).

Valenzuela, G. R. (1978): *Boundary-Layer Meteorol.*, **13**, 61–85.

Wentz, F. J. (1975): *J. Geophys. Res.*, **80**, 3441–3446.

Wentz, F. J. (1992): *IEEE Trans. Geoscience Remote Sensing*, **30**, 960–972.

Wilheit, T., A. T. C. Chang and A. S. Milman (1980): *Boundary-Layer Meteorol.*, **18**, 65–77.

Wilson, B. W. (1965): *Dtsch. Hydrogr. Z.*, **18**, 114–130.

Wright, J. W. (1966): *IEEE Trans. Antennas and Propag.*, **AP-14**, 749–754.

Wright, J. W. (1968): *IEEE Trans. Antennas and Propag.*, **AP-16**, 217–223.

Yoshino, M. *et al.* (eds.) (1985): *Dictionary of Climatology/ Meteorology*, Nimoniya-Shoten, 742 pp. (in Japanese).

Zhao, D. and Y. Toba (2001): *J. Oceanogr.*, **57**, 603–616.

Index

α, 78, 83
$\alpha_{s.t.p}$, 78
absolute vorticity, 80
absorption coefficient, 259
abyssal circulation, 214
accelerated Liebmann method, 202
acceleration potential, 83, 86
acoustic waves, 270
active sensor, 249
ADEOS, 286
advection equation, 196
advection-diffusion process, 114
air–sea–land interaction, 176
altimeter, 250, 286
AMI, 286, 291
Anderson and McCreary model, 176
annual subduction rate, 89
Antarctic Intermediate Water, 100
AOU, 82
AOUR, 93
apparent oxygen utilization, 82
apparent oxygen utilization rate, 93
APT, 254
Arctic Ocean, 225
atmosphere–ocean coupling unstable mode, 187
atmospheric window, 249
AVHRR, 245, 253, 289, 293

β, 83
backward scheme, 200
baroclinic mode, 149
Beaufort scale, 119
black body, 240
Boltzmann constant, 240
boundary layer
   macroscopic structure, 18
Boussinesq approximation, 10

Bragg scattering, 282
Brooke Benjamin, 28
Brunt–Väisälä frequency, 11
bubble entrainment, 52, 270
bulk coefficients, 118
bulk formulas, 48
bulk method, 117
bulk model, 67
bulk temperature, 260
bulk transfer coefficients, 49
Bye, 47

centered difference, 198
Central Mode Water, 99
CFL condition, 199
Charnock's formula, 52
climate jump, 139
climate system, 111
closure problem, 66
COADS, 117
coefficient of viscosity, 7
complex dielectric constant, 273
composite scattering model, 283
conditioning (environmental) parameters, 15
consistency, 199
continuous model, 67
contraction coefficient for salinity, 83
convective adjustment, 204
convective mixing, 64
convergence, 199
core analysis, 84
Coriolis parameter, 80
Coupled General Circulation Model, 233
coupled ice–ocean model, 221
coupled ocean–atmosphere model, 229
Crapper, 29
critical height, 37

$\Delta_{s,t}$, 78
δ, 78
δ_t, 78
daily mixed layer, 63
deep-water wave, 27
delayed oscillator, 180
delayed oscillator equation, 181
diffusion equation, 202
dimensions, 13
  fundamental-, 13
directional distribution function, 36
dispersion relationship, 27
dissolved oxygen, 82
DBBL, 47
DMSP, 278
drag coefficient, 49
Dual-Window method, 262

East Asian monsoon, 90
East Asian wintertime monsoon, 94
East North Pacific Central Water, 75
Eastern Atlantic pattern, 128
eddy
  diffusion coefficient, 8
  diffusivity, 8
  viscosity, 8
  viscosity coefficient, 8, 22
eddy correlation method, 117
eigenvalue problem for coupled instability,
  169
Eighteen Degree Water, 90
Ekman layer, 71
Ekman number, 11
Ekman pumping, 87
El Niño, 5, 132, 144, 212
Empirical Orthogonal Function analysis, 132
empirical orthogonal functions (EOF), 234
energy dissipation, 57
energy input, 57
energy spectrum
  directional-, 36
ENSO, 5, 144, 212
ENSO event, 129
entrainment velocity, 68
equation of conservation, 5
equation of continuity, 6
equation of motion, 6
equatorial Rossby wave, 152
equatorial wave, 151
equilibrium range, 43
equilibrium solution, 182
ERS-1, 291
Eurasian pattern, 128
evaporation, 164

exchange coefficients, 118
external conditions, 15

Feir, 28
finite difference approximation, 197
flux adjustment, 232
forward difference, 198
forward scheme, 200
frequency spectrum, 36
freshwater flux, 124
Fresnel's formula, 273
friction velocity, 19

gas exchange, 55
general circulation, 4
geophysical fluid, 10
geopotential, 85
GEOSAT, 287
geostationary meteorological satellite, 250
geostrophic balance, 11, 85
geostrophic contour, 216
geostrophic current, 75
Gerstner wave, 33
global coupled ocean (ice)–atmosphere model,
  233
global warming, 236
gradient method, 117
gravity-capillary waves, 267
greenhouse effect, 1
group velocity, 31, 57

Hasselmann, 45, 58
HRPT, 254
Huang, 45, 46
hypsometric effect, 217

ice model, 222
in situ density, 78
in situ specific volume, 78
in situ SST, 260
in situ temperature, 77
incipient breaking, 39
inertial frequency, 12
initial wavelet, 265
instability
  Benjamin–Feir-, 28
integral model, 67
Interdecadal variation, 233
International Practical Temperature Scale of
  1968, 77
International Temperature Scale of 1990, 77
inverse method, 105
IPTS-68, 77
ISCCP, 253

isopycnal mixing, 203
isopycnal surface analysis, 83
ITS-90, 77

Jähne, 44
Jones, 52
JONSWAP, 45, 287
JWA3G, 59

Kelvin wave, 152, 155, 211
KEYPS equation, 24
kinematic coefficient of viscosity, 7
Kirchhoff's law, 240
Komori, 57
Kuroshio, 65, 289
Kuroshio Bifurcation Front, 99
Kuroshio Countercurrent, 90
Kuroshio Current, 258
Kuroshio recirculation, 90

La Niña, 130, 145
laminar flow, 17
Langmuir circulation, 71
large-meander path, 96
large-meander period, 90
latent heat flux, 124, 243
lateral induction, 89
law of the wall, 19
leapfrog scheme, 201
level of no motion, 75
Liebmann method, 202
linear wave equation, 196
local equilibrium, 42
log-linear law, 24
logarithmic boundary layer, 21
long-wave radiation, 122, 240
Longuet-Higgins, 32, 34, 43, 46
lower boundary layer of the atmosphere, 164

main thermocline, 87
mass transport, 30
Masuda, 57
Matsuno scheme, 200
McLean, 28
mechanical mixing, 64
method of coupling, 231
microwave radiation from the sea surface, 274
microwave radiometer, 277
microwave scatterometer, 285
middle salinity minimum, 102
Miles, 37
Miles' mechanism, 37
Mitsuyasu, 40, 46
mixed gravity–Rossby wave, 152

mixed layer, 63
modeling of ENSO, 232
modeling of western boundary currents, 205
MOI, 94
momentum flux, 243
momentum flux density, 7
Monahan, 52
monsoon index, 94
Moskowitz, 45
MRI-II, 59
MRI-III, 60
Multi-Channel SST, 263
Munk, 40

Navier–Stokes equation (N–S equation), 7
necessary condition for instability, 169
neutral surface, 83
non-large-meander path, 96
non-large-meander period, 90
nondimensional, 8
nondimensional roughness parameter, 51
nondimensional variables, 15
nondimensionalization, 13
North Atlantic Deep Water, 216
North Atlantic Subpolar Mode Water, 99
North Atlantic Subtropical Mode Water, 90
North Pacific Central Mode Water, 98
North Pacific Intermediate Water, 100, 104
North Pacific Subtropical Mode Water, 90
North Pacific Tropical Water, 100
NSCAT, 286
numerical modeling, 195
nutrient, 82

ocean–atmosphere interaction, 93, 96, 97, 111, 143
ocean–atmosphere-coupled instability, 167
ocean–atmosphere-coupled system, 76
ordered motion, 39

Pacific/North American pattern, 126
Pacific–Japan pattern, 130
parametrization, 48
passive sensor, 249
peak enhancement function, 45
phase speed, 27
Phillips, 37, 43, 44
Phillips spectra, 282
physical constants, 15
Pi theorem, 13
Pierson, 45
Planck's formula, 240
planetary vorticity, 81
Poiseuille law, 17

potential density, 78
potential temperature, 76
potential vorticity, 79
practical salinity units, 77
probability density function of the surface
    slope, 265
psu, 77
pycnostad, 90

radiation flux, 240
radiation stress, 34
radiation-convection process, 114
radiometer, 249
reflection coefficient, 259
regime shift, 139
relative vorticity, 80
relaxation method, 202
relaxation time, 111
resonant scattering model, 280
Reynolds number, 10, 17, 53
Reynolds stress, 8
Reynolds' axiom, 8
RIAM, 57, 60
Richardson method, 202
Richardson number, 23, 66
Rossby number, 11
Rossby wave, 155, 209
roughness length, 23
roughness parameter, 22
roughness Reynolds number, 22

σ, 77
σ_1, 78
σ_3, 78
σ_θ, 78
σ_t, 78
SASS, 246, 285
saturated dissolved oxygen, 82
scatterometer, 250
Schwartz, 28
sea ice, 221
Sea of Japan, 243, 287
sea-surface fluxes, 239
sea-water droplet, 52, 54
Seasat, 245
seasonal mixed layer, 63
semi-empirical, 18
sensible heat flux, 122, 243
shallow salinity minimum, 100
short gravity waves, 267
shortwave radiation, 122, 240
side-band instability, 28
significant wave, 40

similarity law
    Reynolds'-, 10
skin temperature, 260
slab model, 67
SMMR, 277
solar radiation, 113
source function, 57
South Oscillation Index, 144
South Pacific Subtropical Mode Water, 90
Southern Oscillation, 144
specific volume, 78
specific volume anomaly, 78
spectral density, 36
spectral energy density of the surface elevation,
    265
specular reflection model, 279
spin-up, 209
Split-Window method, 262
SSM/I, 278
SSMI, 289
SST (Sea Surface Temperature), 253
stability, 199
stability function, 23
stability length, 24
stability of the delayed oscillator, 183
steepness, 42
Stefan–Boltzmann constant, 241
Stewart, 34
Stokes drift, 31
Stokes wave, 26
Subduction Oscillation, 139
subduction, 88
subduction rate, 88
subtropical gyre, 206
subtropical mode water, 65
Subtropical Mode Water, 75, 82, 90
sun glitter, 266
surface waves
    energy, 30
    momentum, 30
Sverdrup, 40
Sverdrup relationship, 215
SWAMP Group, 59
synoptic, 86
synthetic aperture radar, 250

θ, 76
θ_1, 77
θ_3, 77
θ–S diagram, 79
teleconnection, 126
theory of ventilation, 207
thermal expansion coefficient, 83

thermal inertia, 115
thermal stratification, 23
thermodynamics of the oceanic mixed layer, 162
thermohaline circulation, 75, 205
thermostad, 90
thermosteric anomaly, 78
Thorpe, 55
3/2-power law, 41, 60
Toba spectrum, 282
TOGA, 190
TOHOKU wave model, 58
transmission coefficient, 259
trapezoidal scheme, 200
Triple-Window method, 262
Tropical Water, 100
Tulin, 47
turbulence, 7
turbulence model, 67
turbulent boundary layer
    downward bursting-, 47
turbulent diffusion time, 115
turbulent energy, 66
turbulent flow, 17
turbulent flux, 240, 242
two-scale model, 284
T–S curves, 73
T–S diagram, 73

Uji, 59
upwelling entrainment, 163

validation, 264
velocity potential, 24
velocity profile, 21
ventilated thermocline theory, 87
ventilation, 87, 207
ventilation numerical model, 208
vertical pumping, 89
viscous stress tensor, 7
viscous sublayer, 19
VISSR, 253, 290
von Kármán constant, 21

WAMDI Group, 59
water
    individual-, 39
    unique characteristics, 2
water mass, 73, 74
water type, 73, 74
wave age, 42
wave breaking, 52
    Stokes' limiting criterion, 28
wave breaking parameter, 52
wave current, 31
wave forecasting, 57
wave interactions, 57
wave models, 58
wave momentum, 31
wave steepness, 28
wave–current interactions, 33
wave–wave interactions, 34
waves
    rhombic-, 38
Weddell Sea, 215
West North Pacific Central Water, 73, 75
Western Atlantic pattern, 128
western boundary currents, 205
Western Pacific pattern, 128
whitecap coverage, 52, 54
whitecaps, 269
Wilson, 40
wind drift, 28
wind stress, 49, 120
wind waves, 18, 39
    initial generation, 36
wind-driven circulation, 75, 205
windsea, 39
Wyrtki jet, 157

Yamase, 291
Yanai–Maruyama wave, 152
Yoshida jet, 157

Zakharov, 46
Zebiac and Cane model, 190

Ocean Sciences Research (OSR)

Vol. 1   Coastal Oceanography, Tetsuo YANAGI,1999

ISBN 0-7923-5895-3

Vol. 2   Dynamics and Characterization of Marine Organic
Matter, Edited by N. Handa, E. Tanoue and T. Hama

ISBN 0-7923-6293-4

TERRAPUB, Tokyo / KLUWER, Dordrecht, London, Boston